The Social Contract

ROBERT ARDREY

The Social Contract

A Personal Inquiry
into the Evolutionary Sources
of Order and Disorder

DRAWINGS BY
BERDINE ARDREY

NEW YORK   ATHENEUM

TO THE MEMORY OF

Jean-Jacques Rousseau

Portions of this book were first published in LIFE.

Contents

Bibliographical Key

To avoid the use of footnotes or reference numbers within the text, the key, which follows the text, is provided as a guide to the numbered references in the Bibliography. For each section of each chapter the reader will find appropriate identification of material in the text, along with a number to indicate the source from which it is drawn. When the source is a private communication, it is indicated by PC.

Robert Ardrey

Robert Ardrey was born in Chicago in 1908. He graduated from the University of Chicago, where he began studying the sciences of man. For years a successful playwright, Mr. Ardrey returned to the study of man in 1955 on a visit to Africa. The resulting series of books began with *African Genesis* (1961) and continued with *The Territorial Imperative* (1966) and *The Social Contract* (1970). When not researching in the field, he lives in Rome, Italy, with his wife, the former Berdine Grunewald.

The Social Contract

1. Tuskless in Paradise

A society is a group of unequal beings organized to meet common needs.

In any sexually reproducing species, equality of individuals is a natural impossibility. Inequality must therefore be regarded as the first law of social materials, whether in human or other societies. Equality of opportunity must be regarded among vertebrate species as the second law. Insect societies may include genetically determined castes, but among backboned creatures this cannot be. Every vertebrate born, excepting only in a few rare species, is granted equal opportunity to display his genius or to make a fool out of himself.

While a society of equals—whether baboons or jackdaws, lions or men—is a natural impossibility, a just society is a realizable goal. Since the animal, unlike the human being, is seldom tempted by the pursuit of the impossible, his societies are seldom denied the realizable.

The just society, as I see it, is one in which sufficient order protects members, whatever their diverse endowments, and sufficient disorder provides every individual with full opportunity to develop his genetic endowment, whatever that may be. It is this balance of order and disorder, varying in rigor according to environmental hazard, that I think of as the social contract. And that it is a biological command will become evident, I believe, as we inquire among the species.

Violation of biological command has been the failure of social man. Vertebrates though we may be, we have ignored the law of equal opportunity since civilization's earliest hours. Sexually reproducing beings though we are, we pretend today that the law of inequality does not exist. And enlightened though we may be, while we pursue the unattainable we make impossible the realizable.

2

The propositions that I have put forward are not self-evident. Were they so, then there would be no need for me to write this book, or cause for anyone to read it. Neither do I put them forth as subjects for immediate acceptance or rejection, immediate digestion or expectation of instant nourishment. Indeed, like uncooked rice, they are quite indigestible. But in the course of this inquiry you and I will do a bit of cooking and see what comes of it. And so, to begin with, I suggest that we content ourselves with simply lighting our fire. Let us inspect the dream that has brought to the climactic years of the twentieth century the assurances and rewards of a madhouse.

The philosophy of the impossible has been the dominant motive in human affairs for the past two centuries. We have pursued the mastery of nature as if we ourselves were not a portion of that nature. We have boasted of our command over our physical environment while we ourselves have done our urgent best to destroy it. And we have pursued the image of human equality as citizens of earlier centuries pursued the Holy Grail.

The grand escapade of contemporary man can be denied neither excitement nor accomplishment. Out of our dream of equality we have lifted masses from subjection, moved larger masses into slavery. We have provided new heroes, new myths, new gallantries; new despots, new prisons, new atrocities. Substituting new gods for old, we have dedicated new altars, composed new anthems, arranged new rituals, pronounced new blessings, invoked new curses, erected new gallows for disbelievers. We have reduced sciences to cults, honest men to public liars. We have even reduced the eighteenth-century vision of human equality, glorious if false, to a more workable twentieth-century interpretation, mediocrity, inglorious if real.

Fundamental though the natural impossibility of equal beings must be to this inquiry, still it is not all. And if we are to glimpse a social contract leading neither to tyranny nor to chaos, then I prefer at first to consider it simply as a fraction of a larger delusion. The philosophy of the impossible rests on an article of faith, that man is sovereign. And the Greeks had a word for it: hubris.

To lift your head too high: it is to challenge the gods and risk a few thunderbolts assaulting your skull. The skeptical Greeks, never excessively infatuated with gods themselves, turned less to supernatural than to natural explanations of why the world is the way it is. Yet never, from the early times of the Ionian philosophers down through later excursions and controversies of the lively Greek mind, am I aware of presumptions that man could master nature. Even Protagoras' celebrated statement that man is the measure of all things seems to have been intended more in praise of the individual than in denial of forces larger than man.

As in Western thought various tides have swept this bay, assaulted that promontory, or, receding, have bared undistinguished flats, so we have turned now to gods, now to God, again to nature and its laws for satisfying answers. Our postures have varied from the compliance of slaves to the confidence of sailors. At our best we sought solutions of relevance to man; at our worst we avoided them. But never, till modern times, did men in any significant number presume a human sovereignty much larger than the human shadow. Never did we risk the Greek hubris, and a shattering knock on the head. "The conquest of outer space" for a most inquiring Greek would as a phrase have seemed as dangerous a possibility as it remains, in all fragility, a phrase of small reality today.

The big brag preceded the big bang as a human possibility. Any demonstration that the earth revolves about the sun, while offensive to authorities in charge, did not presume that we could reverse its course. Any proof of a natural law called gravity did not presuppose that man could make apples fall up; designers of supersonic planes, indeed, still take account of the apple. To the frontiersmen of science the discovery of natural laws meant no more than that we had explored certain forces governing the dispositions of man. But for many a hoi-polloi scientific settler who came after the frontier such discoveries meant something quite different. Man could master nature.

Eighteenth-century rationalism, while dispensing with the supernatural as a governing force, left a vacuum that not all the Encyclopedists could fill. And so an alliance between nineteenth-century optimism, looking to the perfectibility of man, and the early modern scientists (Darwin was not among them) rushed in. And the sovereign rule of materialism came about. Man, with the aid of science, could do anything. As materialist were the socialist philosophies as the capitalist. Uninhibited by laws natural or divine, we busied ourselves with the building of Paradise.

And no mean thing is this Paradise of the Impossible. Could animals dream, then our material heaven might well be the stuff that their dreams are made of. The small-brained hominid, dragging himself through the millions of years of our evolution, may well have longed for supermarkets. Yet he, I suspect, even facing the hostile African night, had a sense of certainty. And we have none.

A philosophy of the impossible is indeed no philosophy at all. And a paradise lacking a philosophy is one of uncertain future. Aimlessly we prowl our highways, teach or attend our classes, swallow our drugs or our television dinners, quarrel, fornicate, fear our children, sigh for the unfortunate and avoid their presence, envy the fortunate and court their approval, work to forget our meaningless lives, drink to forget our meaningless work, purchase our pistols, deplore all wars, and praise the dignity of man.

It cannot be said that man, installed in his self-made heaven, has lost his dignity. The buffalo, small of brain, peering out of

the African bush, commands dignity. Nor can it be said that if
we have made mistakes we cannot learn. The amoeba can learn.
Back in the 1920's an experiment was arranged in a darkened room whereby an intense beam of light barred the movement of amoebae. Among the brainless students there was one
who never learned. On trial after trial it persisted in its effort
to cross the beam of light. But there was one who tried just five
times and never moved in that direction again. Not only organisms lacking the least brain or nervous system could learn,
but, significantly there was wide variation in their gifts.

We have our dignity, which is the dignity of living beings. If
we have made errors, then—since an amoeba can learn—there

must be among us those likewise gifted. But what is it we must
learn? In another time we should have taken our troubles to
the priest in a certainty of faith. But the faith is gone and the
priests are missing. Man—omniscient, omnipotent man—has
none to talk to but himself. And, worst of all, it is how we
wanted it.

Who will save us? Who will inform us? We turn to science,
our sole religion, our one maker of miracles. It is science that
adjudicates the rivalries of nations, dictates economic triumphs,

decrees disasters. It is science that with cosmic disregard for human fate adjusts the balance of military terror now this way, now that. It is science that saves lives here, destroys them there, perfects new means of postponing the grave, new means of making life unendurable. It is science that with perfect casting has assumed the role of the Unknown God. Yet were I the scientist—not science but the breathing, aching scientist who suffers from indigestion and achieves the respect of all but his children—and were I asked to save us, then I think I should put on a false mustache and other appropriate disguise, go out the back door, and vanish like some extinct bird over a former horizon.

Yet it is not quite so. For what God has granted, God may take away. And it lies within the power of that present god, the individual scientist, to withdraw from mankind the illusion of sovereignty that science, in partnership with obsolete philosophies, has created. But courage as much as competence must be the endowment of such a rebel god.

Natural law has been variously defined, as it has been variously abused. It might be described, in contrast with civil law, as the kind of law you discover only after you have broken it. Such is the predicament of contemporary *Homo sapiens*, who, looking about at his program of disaster, asks, "What did I do?" His refuge may lie in social paranoia such as that so favored by the young. It is somebody else's fault. But the mature must inquire more deeply. What did we do that was wrong? And there is coming about in our time a generation of scientists who, granted the courage, have the power to answer.

A natural law is one made not by state, not by religious authority, not by man at all, but one which human reason may investigate, recognize, and prove. "Natural law" has been invoked by many—by kings, for example, to support their right to rule—but without proof. The ease with which "natural law" has been enlisted to invest with sanctity many a position of the *status quo* has given the term a bad name in our time. Yet natural laws exist. And they lie beyond human power to veto or amend. Unlike other laws, they deal impartially with big-brained man or small-brained African buffalo. And it should be science that can identify them for us.

Yet, to the bewilderment of the layman, scientists, as we shall see, do not provide the same answers. There is nothing new about passionate divisions within an accepted religion. But these are troublesome times for the sciences. A generation of paleontologists—Raymond Dart, Robert Broom, L. S. B. Leakey —has demonstrated that man's evolution from some gentle, ancestral forest ape is not what it was thought to be in Darwin's time. A generation of ethologists, pioneered by Konrad Lorenz, Niko Tinbergen, C. R. Carpenter, has shown the behavior of animals in a state of nature to be not at all as we presumed. A generation of population geneticists—Sewall Wright, J. B. S. Haldane, Sir Ronald Fisher—has altered beyond recognition our former concepts of heredity. A generation of biologists, among them Sir Julian Huxley, George Gaylord Simpson, Ernst Mayr, has synthesized the advances of this century with older Darwinian theory to produce a new biology so revolutionary as to remain beyond the grasp even of some of its contributors.

Few members of the natural sciences, however, would today dispute the evolutionary continuity of man and the natural world, or uphold the proposition that for man exists one fate, for nature another. Difficult it may be to part with the convictions of two centuries, and for some it is impossible. Yet part with them we must, for while the conviction of human sovereignty has led us to dare and aspire, it has led us likewise into the Age of Anxiety. We build paradises in which we have no faith.

When we renounce our hubris; when we see ourselves as a portion of something far older, far larger than are we; when we discover nature as our partner, not our slave, and laws applying to us as applying to all: then we shall find our faith returning. We have rational faculties of enormous order. We have powers granted never before to living beings. But we shall free those powers to effect human solutions of justice and permanence only when we renounce our arrogance over nature and accept the philosophy of the possible.

3

Since we are inspecting man as a portion of nature, the capacity for lying should not be skipped. A few students of human language have implied that only through the complexity of our communication has the telling of lies become possible. I seize on the happy opportunity to announce that lying is a natural process. Man has enough to answer for; he need not answer for this.

Some of the most outrageous liars in the natural world are found among species of orchids. To gain perspective on what might be called natural square deals as opposed to natural larceny, we may recall that plants evolved before birds and flying insects, and depended on wind or water to scatter their pollen. It was an inefficient system, and a colorless one, too, in this time before flowers. But then came the insects and sensuality became possible. The scents we enjoy, the colors we delight in, evolved as signals to attract this insect, that bird. Partnerships were established—what zoologists call symbiosis—so that the fuchsia, for example, offered the hummingbird nectar in exchange for hauling fuchsia pollen around. Everyone got a fair shake. But there must always be liars.

There are species of orchids that have puzzled naturalists since the days of Darwin. They seemed to offer no inducement as their part of the deal, yet still insects did their job. At last, in 1928, a woman named Edith Coleman solved the problem in her study of an Australian orchid named *Cryptosylia*. The scent, a perfect imitation of the smell of the female of a species of fly, acted as an aphrodisiac on the male. He was drawn to the flower. There he encountered as part of the orchid's structure a perfect imitation of the female's abdomen. It was all too much for him, and in his efforts to copulate with the orchid he got himself nicely dusted with pollen. I am aware of no more immoderate fraud in the natural world.

Fraud, however, is normal in nature. There are deep-sea fish prowling dark depths with lanterns on their snouts. Smaller fish are attracted by the light and promptly eaten. There is a poisonous Ceylon snake resembling the viper with a brightly colored tail-tip which can be wriggled like a worm; lizards have been observed biting the tail and immediately being struck by the fangs at the other end of the living trap. Snakes, indeed, seem to deserve their long-established reputation for the fascination which they exert on their prey. A Madagascar snake called *Langaha nasuta* has a weird structure on its head resembling a finger, which it slowly moves as it approaches its victim. The victim normally stares too long. A South American snake, *Oxybelis*, has achieved total mastery of the unacceptable. It is a tree snake, and except for a vivid neck, its body resembles a vine. The top is olive, the underside cream, and on its side is a black, horizontal line. The tongue has precisely the same color scheme, and when fully extended from its slender snout seems a part of the body. Prepared now to be fascinating, *Oxybelis* begins to expand and contract its vivid neck, achieving in the process resemblance to nothing on this earth. Reliable observers have reported lizards hesitating, approaching—the only word is anthropomorphic—perplexed. And struck.

Not all the wonders of natural fraud are the property of villains, however. Many a lie is told on behalf of the potential victim. The tropical fish called *Chaetodon*, for example, has spots resembling eyes on either side of its tail. It swims slowly backwards, apparently head-on. But if a predator strikes at it, the fish is off at high speed in the proper direction. All camouflage, indeed, is deception. That both fish and seabirds tend to have white undersides to provide camouflage against the sky from an underwater point of view has been long assumed. The proposition was demonstrated during World War II when British planes on antisubmarine patrol improved their records by

painting the planes' undersides white. It is further confirmed by a mixed-up creature, the Nile catfish, who through some unhappy mutation got his white on top and the dark beneath. He compensates successfully by swimming upside down.

In terms of duplicity, the Ceylon shrike has perfected a lie of utmost complexity. The bird's full name is the Ceylon black-backed pied shrike, and its ways of nesting were described almost thirty years ago by W. W. A. Phillips in the ornithologists' journal, *Ibis*. As a site for their nest the parents choose that most conspicuous position, the bare open fork of a branch. Here they build a tiny cup of fiber, plastered all about its outside wall with lichens bound by cobwebs. At a distance of a very few feet the two-inch nest cannot be distinguished from a knot in the branch. This, however, is only the foundation for prevarication.

In the Ceylon shrike one finds a perfect evolutionary union of body, culture, and behavior. The parents are black and white. The young are a mottled color blending precisely with the appearance of the lichen-plastered nest. But it is the behavior of the young that leaves the observer in awe. There are usually three, and when the parents leave the nest the young sit facing the center, immobile, their beaks raised at a sharp angle and almost touching in the center. The tableau presents the most exquisite imitation of old splinters at a break in the branch, and the young will not stir until the parents return. What the family has achieved, and what must puzzle the evolutionist, is a social lie in which each member plays its part.

The natural history of prevarication is, indeed, without end. Human communication, like most of our capacities, has merely provided superb elaboration on an old, old theme. Through our use of words we delude each other with grave conviction; we have our way, as the Ceylon shrike has his. In one sense only is our capacity unique and entirely our own. Man, so far as I know, is the only animal capable of lying to himself.

That we lie successfully to each other is natural; that we successfully lie to ourselves is a natural wonder. And that the three sciences central to human understanding—psychology, anthropology, and sociology—successfully and continually lie to themselves, lie to each other, lie to their students, and lie to the public at large, must constitute a paramount wonder of a scientific century. Were their condition generally known, they would be classified as public drunks.

In another age we might amuse ourselves with the miraculous contradictions inherent in even the most educated of human beings. But this is not such an age. Our century plays out its dark charades. The human outcome approaches like a thundering, unlighted munitions train. In unease we go to bed at night, in unease we rise in the morning. The newspaper carries an item that "science says," and we feel a reassurance. Our children go off to school, and we praise God for education, our last, best hope. They, if the world lasts long enough, will be better able to handle the mess than were we when we arranged it. But will they? Or will they be more successfully brainwashed? We proceed to the office with a calm reassured by the scientific mystique. And the question must of course remain: whose blame is more profound, that of the scientist who lies to himself, or of the responsible citizen who fails to inquire?

I deal harshly with the central sciences of human understanding, and I shall deal still more harshly as this inquiry progresses. Theirs is the responsibility for reconciling man and man, and of providing humanity at large with the accommodations so singularly lacking. Yet the indictment must be placed not just against the few but against all of the scientific community; a temple psychology of sorts invests it. I have spoken of science as our only religion offering an avenue of faith. But scientist is reluctant to speak out against scientist. Temple psychology, as in any religion, prevails. Like a priest conducting his mass in Latin, he presents his conclusions in a jargon that the most intelligent layman cannot translate and thus most unhappily cannot question. Like a participant in some tribal ritual, the scientist conceals his personal identity behind the stylized mask of his trade. The novice priest, taking his hard-earned Ph.D., accepts with his degree the mysteries of the temple. He will be moved by controversy, but he will address only his fellows. He will perfect the dialect of his discipline, frequently unintelligible even to members of the next discipline and certainly to the layman. He will maintain the mystique of infallibility or suffer excommunication: he will get no faculty appointment.

I believe that the publication of James D. Watson's *The Double Helix* in 1968 offers excellent confirmation of how seldom are temple vows broken. Had Watson not received the Nobel Prize for his part in the definition of the DNA molecule, I doubt that the book would have been published. His own university press refused it. When the book at last appeared, the uproar was such as to make it an immediate best seller. But I find myself unconvinced that a book about the DNA discovery would in itself have produced such a sensation. What Watson, however, with such admirable clarity and courage, had done was to tell the inside story, to describe the fallibilities, the jealousies, the overwhelming humanity of the scientist. It was a story seldom told.

Molecular biology lies far from the humanistic sciences that I indict. Yet all scientists must accept the responsibility for hiding from public view, so that scientific infallibility may be preserved, the picture that so many know so well. It is the picture of cultural anthropology, behaviorist psychology, and environmentalist sociology like three drunken friends leaning against a lamppost in the enchantment of euphoria, all convinced that they are holding up the eternal light when in truth they hold up nothing but each other.

Each is no more than a school of science, a division of the discipline each dominates. All hold to a central assumption that the human brain owes little or nothing to evolutionary experience. The sociologist since the days of Durkheim denies biological influence on our social arrangements, and in a bastard paraphrase of Protagoras maintains that the proper study of sociology is sociology. The behaviorist maintains that all of human behavior is predicated by the conditioned reflex, that the baby born comes into this world a *tabula rasa*, with neither genetic prejudice nor inborn identity, and will act as did Pavlov's salivating dogs in predictable response to association of punishment or reward. Man is clay; no more. Harvard's B. F. Skinner, with his reinforcement theory, is today the czar of such psychology. Cultural anthropology, the third of the sciences, maintains that man is a product solely of his culture. Why the young of our time revolt against the culture that presumably created them is a question for which our anthropologists provide no direct answer.

Yet these three sciences of human understanding dictate the education of your children and mine, since they dictate who will receive a Ph.D. and become our children's teacher. A true establishment exists, a mutual-aid society like the three drunks under the lamppost. B. F. Skinner may turn his attack on Europe's greatest student of animal behavior: "Konrad Lorenz' *On Aggression* could be seriously misleading if it diverts our attention from manipulable variables in the current environment to phylogenetic contingencies which, in their sheer remoteness, encourage a nothing-can-be-done-about-it attitude." For the reader unable to cut through the jargon, I give my word that the psychologist Skinner is defending environmentalist sociology. Our distinguished cultural anthropologist, Ashley Montagu, in an attack on one of my own books, makes the remarkable statement: "The notable thing about human behavior is that it is learned. Everything a human being does as such he has had to learn from other human beings." Professor Montagu is coming to the aid of Professor Skinner.

In its qualities of reciprocal reinforcement, the mutual-aid society seems all but invulnerable. Dissenters exist. Harry F. Harlow, professor of psychology at the University of Wisconsin, may demonstrate beyond argument that exploration, the innate drive to learn, comes in every baby born as well as every animal package. Yet the learning theory of reward and punishment carries on. Zoologists at the University of California at Los Angeles may demonstrate beyond argument that without reward or threat of punishment, little wild rodents will learn in such measure as to dazzle the human being. Skinner prevails. Jerry Hirsch, psychologist at the University of Illinois, will attack Skinner's conclusions concerning laboratory rats as the consequence of using inbred domesticated animals for experiments irrelevant to the human being. Skinnerism replies with one of the most remarkable comments in supposedly scientific literature: "We are primarily interested in the most domesticated of all animals—man." Yet throughout all the natural sciences the definition of a domesticated animal is one that is the product of controlled breeding, which man—aside from a few temporary and unsuccessful efforts in the periods of slavery—is not.

This is not science. Then what is it, when the layman reads, "Science says"? Or his child, persuaded, learns as science teaches? It is the inheritance today of the nineteenth-century devotions of Bishop Wilberforce throughout his debates with Darwin's disciple, T. H. Huxley. It is the defense of man, the unique being, carried through a myriad of rationalizations. It is anti-evolutionism, minus divine special creation.

The rationalizations vary. There is the appeal to culture, such as was made in 1960 by Marshall Sahlins, one of the younger leaders of cultural anthropology. Choosing as his text Sir Solly Zuckerman's conclusion published in 1932, itself based entirely on the behavior of captive baboons in the London Zoo, that primate society is founded on sexual attraction, Sahlins demonstrated that human society is not. In a notable issue of the *Scientific American* (September 1960) he presented as Scripture:

> There is a quantum difference, at points a complete opposition, between even the most rudimentary human society and the most advanced subhuman primate one. The discontinuity implies that the emergence of human society required some suppression, rather than a direct expression, of man's primate nature. Human social life is culturally, not biologically, determined.

Sahlins' Scripture lasted just one year. In 1961 S. L. Washburn and Irven De Vore published in the same *Scientific American* their now-classic study of baboons in the wild. Washburn is professor of anthropology at Berkeley, and though on rare occasion I may argue with him, I regard him as our greatest anthropologist. De Vore was his student and today is a professor at Harvard. For ten months they had watched baboon troops in East Africa. "Our data offers little support for the theory that sexuality provides the primary bond of the primate troop." They quote Sahlins on the sexual magnet. "Our observations lead us to assign to sexuality a much lesser, even at times a contrary, role."

Throughout the 1960's massive studies of primate societies, to which we shall continually be turning in this investigation, offered confirmation of the Washburn-De Vore conclusion. J. J. Petter's studies of various lemur species in Madagascar showed short sexual seasons, yet tight all-year societies. The same was found true of Japanese and rhesus monkeys. George Schaller's mountain gorillas, in bands as large as twenty-seven under the absolute control of a single dominant male, showed low sexual activity. In 466 hours of observation Schaller watched copulation twice, both times on the part of subordinate males in the presence of the bored leader. By 1965 Jane B. Lancaster and Richard B. Lee could survey fourteen primate populations in a state of nature and conclude, "It is clear that constant sexual attraction cannot be the basis for persistent social groupings of primates." The following year the Harlows wrote of "the demise of the sex theory."

And what in the meantime had happened to the discontinuity between human and animal societies, and to the corollary that "human social life is culturally, not biologically, determined"? Nothing. Sociology and cultural anthropology carried on as if nothing at all had happened. And with equal poise they ignored the inroads of linguistics.

A rationalization comparable to that of Sahlins rests on language and the human capacity for verbal communication as evidence for cultural independence from our evolutionary background. But expanding studies of animal communication have reinforced the revolutionary conclusions of such students of linguistics as Noam Chomsky and Eric Lenneberg that a child's rapid learning of language could not be possible if biological patterns were not as much present as those motor patterns making possible walking. Even our unique capacities for speech are placed among the characters that have come to us through the evolutionary way.

Masses of hard evidence are today destroying the essential premise of the three central sciences of human understanding, that a discontinuity exists between human and other animals. I might until recent years have accepted a single human capacity as uniquely ours, shared with no creature below the rank of *Homo*. This is our recognition of death and our tendency for ritual. In *African Genesis*, while stating my suspicion that students who subscribe to animal limitation usually turn out wrong, I still accepted the prevailing opinion that no animal recognizes death. Then in later years I encountered, again and again, the Elephant Story. And I turned out wrong.

It is a remarkable fact that the four animals looming largest in human consciousness—the lion, the wolf, the bear, and the elephant—remain largely unstudied. George Schaller's three-year observation of the lion in East Africa, recently completed and to be published in the next year or so, should take its place as the most important and exciting study ever made of an animal species in the wild. But our information on the wolf is spotty, on the bear all but nonexistent, and on that unfathomable beast, the elephant, just sufficient to make one yearn for the time when somebody with superhuman strength and ingenuity comes to know him the way Tinbergen knows herring gulls.

Two scientists, Richard Laws of Cambridge and Irven Buss of Washington State University, are generally accepted as our ranking authorities on the vast gray giants. Both, however, are

ecologists and have devoted their principal efforts to observation not of elephant behavior but of the elephant's disregard for niceties of relationship to his environment. But in 1963 I ran into Buss in western Uganda during the course of one of his tours of duty, and for the first time heard the Elephant Story. Built somewhat on the proportions of an elephant himself, Buss would command authority even were it not for his scientific reputation.

At the time, the American scientist was experimenting with means to map the movements of elephant herds. It is the capacity of the elephant, despite his size, to vanish into clear equatorial air that makes study so difficult. During a later season, accompanied by a photographic party, I encountered a herd of sixty in the bottom of Tanzania's Rift Valley. It was too late for picture-taking, and so we sat in the midst of the revolving mass, making ourselves as inconspicuous as possible, while we planned photography with the morning's first light. But by morning they were gone. During the night they had climbed the 2,000-foot wall of the valley and vanished onto the escarpment.

It was the kind of problem Buss faced. His tentative solution was to tranquilize a member of an elephant party, attach a radio transmitter to the creature, then follow the beeps with a light plane. Ingenious though the solution might seem, there were two difficulties. The use of a dart gun to inject a tranquilizer into a wild animal was a technique then new, and just how much drug to use on a creature so huge was still a matter of guesswork. And the second was the problem of social defense. We shall discover later in this inquiry that defense is normally the first function of any society, human or other animal. Elephants are among those who have perfected it. And one of Buss's first experiments was a disaster.

What we think of as an elephant herd is normally a family party of several mature cows and their offspring. Buss selected a young cow, her calf close at her side, as his target. But no sooner had the dart penetrated her hide than he realized that he had overestimated the dose. She collapsed like a punctured, withered balloon. He and his African helpers were equipped with antidote for such emergencies, but they could not approach. Confronted by the angry, trumpeting phalanx of defenders of the downed cow, they could only wait and hope. After an hour and forty-five minutes she died.

It was not Buss who first recognized her death. It was her elephant family. In an instant they ceased their defense, moved aimlessly about. Only after that instant did Buss realize what had happened. But it was not the end of the Elephant Story. The oldest cow, perhaps the grandmother of the lot, moved the party away into the edge of the forest. The little calf still lingered by its dead mother. The old one returned, played with it a bit with her trunk, then coaxed it off to join the waiting group. Now the old one returned again. She broke down branches, pulled up grass, and covered the forequarters and the head of the departed. Then she returned to the family, and all, in elephant silence, vanished into the forest.

Recognition of death was inarguable. But was the ritual that followed simply the inexplicable response of a single, strange, inexplicable old female? I told the story at a drinking party on a ship bound from Naples to New York. A White Father—a Canadian and a member of the order that from earliest times in Uganda has had its missions in far African corners—was stunned. Seventeen years earlier, fresh from Montreal, he had been called by a runner to give the last rites to an African poacher trampled by an elephant in the bush. When he arrived, too late, the man was dead. But the corpse had been covered by grass and branches.

In his *The Deer and the Tiger* George Schaller describes an incident in India (and here we deal with a different elephant species) when he staked out a buffalo as bait for tigers. The mother tiger killed it, retired while her cubs ate. Then an elephant appeared while the last cub, frightened, made off. The elephant pulled down branches, covering the remains.

Perhaps the eeriest versions of the Elephant Story are told by George Adamson, the famous "George" of his wife's Elsa books. For a generation Adamson was senior game warden of Kenya's enormous NFD, the Northern Federated District, inhabited by a few diverse tribesmen and a shocking collection of animals. I have no friend who knows more about animals in their natural ways, or whose varied experiences are recalled with such objectivity. Adamson recalls, for example, a native woman on the long walk home who collapsed beside the road, exhausted. She woke in terror to find herself surrounded by elephants feeling her with their delicate trunks. While frozen she lay, they covered her with branches.

The most haunting of Adamson's recollections, however, concerns an elephant he shot in the NFD's little administrative capital, Isiola. A distraught woman neighbor called him to report that an elephant was in her rose garden, demolishing it on the way to the fruit trees. It was part of Adamson's job; so he sighed, got his gun, arrived and killed the beast. The gunfall of meat was turned over to the locals, and the remainder of the carcass, after human satiation, was loaded onto a lorry to be dumped some miles from town. There hyenas would clear up the rubbish. This they did. But some days later elephants appeared in the demolished rose garden. One carried the shoulder bone of the deceased. The shoulder bone was deposited precisely where Adamson had killed it.

Ten years ago all such stories would have been dismissed as hunters' anecdotes. They can no longer be so dismissed. Nor can those defenders of the last stand of the philosophy of human uniqueness ignore indefinitely the mysterious legacies that lie within us.

4

The educated, concerned reader may wonder why, a century after Darwin's *Descent of Man*, other educated, concerned citizens should so resist the human implications of evolution. There must be many reasons, although among them I do not believe that today religious offense is significant. In the academic community, perhaps, one finds frequently those professionals who do not care to admit that they were educated too soon or that they have failed to keep up on their homework concerning contemporary advances in biology. And particularly among those who regard themselves as intellectuals, and who regard the life of the mind as all, there must be not a few who understandably look with distaste and humiliation on conclusions that the mind is not all. Yet I have come to believe that the principal resistance comes from those who through the years have discarded an honorable liberalism in favor of an unrelated, dishonorable dogma, and who conceal beneath a many-colored cloak of humanitarianism the darker, danker garb of self-righteousness.

I know of no more revealing passage from what passes for science than one drawn from Sol Tax's introduction to *Horizons of Anthropology*, a group of essays by some of our younger and more vigorous students. Tax himself is a professor at the University of Chicago, editor of our most readable journal, *Current Anthropology*, and a dominant, quite formidable figure in the field. He writes what is far more a political creed than a scientific estimate, and hubris attracts distant thunderbolts:

> Whether we are archaeologists or linguists, students of the arts or of geography, whether we study the behavior of baboons or the refinements of the human mind, we all call ourselves anthropologists. It will become evident also that we all carry within us the liberal tradition of the first ancestors. Humankind is one: we value all peoples and all cultures; we abhor any kind of prejudice against peoples, and the use of power for the domination of one nation by another. We believe in the self-determination of free peoples. We particularly abhor the misuse by bigots or politicians of any of our knowledge. As scientists we never know all the truth; we must grope and probe and ever learn; but we know infinitely more than the glib racists—whether in the United States or South Africa. We are equalitarians, not because we can prove absolute equality, but because we know absolutely that whatever differences there may be among large populations have no significance for the policies of nations. This comes from our knowledge as anthropologists: but it also pleases us as citizens of the world.

These are not the words of a free mind, nor do they express a proper discipline for a free science. That the first sentence of this chapter will press all the predictable buttons in Professor Tax, provoke all the loaded, stereotyped phrases of invective so characteristic of animal communication, is regrettable. My regrets, however, are bearable. In an era when a sense of catastrophe invests much of humankind, then anthropology, the science of man, must have more to offer than self-congratulations.

That not all of Tax's colleagues agree with him is evident even within the book to which he writes his introduction. The most severe reproach to his school of scientific thought, however, was delivered several years later by Julian H. Steward of the University of Illinois, like Tax a cultural anthropologist of high authority but with more profound intimations as to anthropology's obligations. In a letter to the journal *Science*, Steward wrote:

> To those who claim that the social scientist cannot separate his science from his human compassion, I answer that he can and he must. . . . It should be the task of the social scientist to develop a methodology that will permit predictive hypotheses rather than to make moral exhortations.

The obligations of the sciences of human understanding to human welfare and reconciliation are such that without them their disciplines would have no reason to exist. Such an obligation dictates both humility and objectivity. It will not do to be pleased with oneself as a citizen of the world when the world is in trouble.

The demand for hypotheses of predictive value inspires this work, as it has inspired my previous investigations. If accepted doctrines of human sovereignty, uniqueness, perfectibility had provided such hypotheses, then we should be all right. But I find it difficult to believe that this century would have left quite such an irreproachable record of massacre and terror, of high intentions frustrated and low intentions consummated, had we been guided by other than error; just as I cannot believe that this century's legacy to the next should include quite so many problems without answers. We have given the philosophy of the impossible its try. Is it permissible to suggest that we turn to another?

That a hypothesis derived from study of man's evolutionary nature may have predictive value receives a certain confirmation from the lamented Vietnam war. In *The Territorial Imperative* I inspected the history and the nature of a biological force called territory, first diagnosed in bird life by the British amateur ornithologist Eliot Howard. Today we can predict that in many species other than birds the male will defend his territory—his exclusive bit of the world's space—against all intrusion with a high probability of success even though the intruder be the stronger. A corollary to the proposition I termed the amity-enmity complex, the likelihood that a group of defenders of a territory will be drawn together, united, and their efforts compounded by the intruding enemy. I suggested that man is a territorial species, and that we defend our homes or our homelands for biological reasons, not because we choose but because we must.

I was writing early in 1966, when escalation of American power in Vietnam was less than a year old and American optimism was still a native resource. Applying the territorial principle, I published my conclusion that the war was unwinnable. A powerful intruder, uninhibited by world censure, may with a single blow annihilate a territorial defender. But an effort to escape moral obloquy through gradual escalation of force gives the weakest defender opportunity to escalate his own quite incalculable biological resources. Incapable of playing the Hitler, we played instead evolution's fool.

There is an ironic footnote to America's most profound humiliation. Any insight into man's evolutionary nature would have determined that the war must not take place. It was belief in man's *rational* nature—acceptance of psychology's reinforcement theory that, confronted by sufficient punishment, tempted by sufficient reward, men will come to their senses—that encouraged hope in our escalation policy. And in utmost humility we must add a footnote to our footnote. The final decision to escalate and that ultimate of follies, to bomb North Vietnam—was made not by the much abused generals, but by a group of the most acute, most educated civilian minds of which my country could boast. And educated they were.

The Vietnam war was the most costly experiment ever designed to test a scientific theory accepted by the most educated of men. And the Vietnamese, in some wild, wild fashion, differed, as events were to demonstrate, from B. F. Skinner's domesticated rats. There is a jungle in the human heart that denies, as so frequently occurs in the jungle of animals, the thesis that might makes right. The territorial principle is a portion of that denial. While in many a minor fashion we may accept reinforcement theory, and the general principle that men confronted by choice will accept the more pleasurable, the less painful, still a breaking point between the reasonable and the unreasonable must be reached. There is too ancient wisdom in natural law to leave the survival of populations, the possession of a territory, to the vagaries of somebody's reinforcement theory. Far too immense is the balance of evolutionary forces to endure such a rational conclusion as that might makes right. And so a weary time in American history came about when we abandoned the question, *When will the Vietnamese learn?* Yet we still have not accepted the question, *When shall we?*

I have reviewed the American maladventure in Southeast Asia not for its substance but to probe the predictive values of hypotheses relating to our evolutionary legacies, in contrast to hypotheses predicated only on human uniqueness. *Homo sapiens* is not a rational being, and were he so, then history would require endless rewriting. But as we possess powers of reason denied other animals, so have we the power of reason to investigate our own irrationalities and their sources. And commanding them, we possess that ancient wisdom, *Know thyself.* It is hubris only that denies our powers.

The dusk of the twentieth century spreads: through our city streets, along our highways, into our factory parking lots. It enters our gardens, our hallways, our administrative offices of industry and state. Are these our children? They seem unfamiliar. Are these our parents? We do not know them. The dusk

34

falls deeper, and it does not discriminate between rich and poor, educated and illiterate, Soviet, Chinese, Frenchman, American. The races themselves become indistinguishable, and we must peer most closely if we are to see each other at all. We must hold the printed word rather close to our eyes, and even then an unnatural shadow obscures its meaning. Yet this dusk is a strange one, an abnormal one, for the sun, if we could see it, stands at noon.

The clock ticks. Never was there such a night as that which threatens. For it is a night of our own fabrication.

5

Man comes before the precipice. We are not unaware. The dark is starless. We wander, lost, and, like an old, old dog, fear is our one remaining companion. Which way is north, which south? Which way is the highland, which way the abyss? We do not know. We search for a light: there is none. We stretch out a hand: the night receives it. We proceed, since procession is of our nature. But which way? And how close is the precipice?

Three billion years of organic evolution lie behind us. It is a respectable age, almost a third as long as that of the universe itself. Three billion years of circumstance and chance, of infinite trial and infinite error, of forgotten failures and compounding successes, have arranged the being called man. The errors, the failures, we may forget with brief mourning. These were not of our ancestry. They were tried and found wanting, condemned to extinction. Shall it be told on the record of living beings that after three billion years the supreme experiment, man, was found wanting?

There is truly nothing new about extinction, excepting only the awareness. History, a great scientist once commented, is a charnelhouse of species. And perhaps the true uniqueness of man arises from his being the first tragic animal. We are the first, so far as we know, permitted self-estimate, permitted remembrance of things long past, permitted visions of things to come. We have been permitted pleasures transcending the joys of other animals, since we may savor ours in time's long contrasts. We have been granted, in full freedom, that exquisite perception revealing the moment that is now as the moment that has never been before. And will never come again.

It is the final freedom that we are granted knowledge of the precipice, even as we have been allowed to construct it ourselves.

In the early years of our wonderings there lived a tragic poet. And he unfolded the drama of a king, even then an old-time figure of remembrance, and it was a tale of such resonance as to be told and retold through the ages since. The king, it seems, had in all innocence offended the gods. And when the truth at last came to him, in horror he put out his own eyes.

Is this indeed an incalculable night through which we fumble our uncertain way? Or is it blindness self-inflicted, are we like Oedipus, have we put out our eyes? Did we in our innocence—and innocence it was—offend certain secret, omnipotent gods? Do we now in our terror of consequence refuse to look? As we constructed the precipice, have we constructed the dark as well?

It is not impossible.

Our foot slips. We fall, *Homo sapiens,* in perfect ignominy on our belly, we feel, reach, search for solid earth. And somewhere in the black, depthless void there is laughter. We rise a little, on hands and knees, peering into the impenetrable. The laughter falls away into derisive chuckles. But something out there is watching us. There is something that can see, while we cannot.

Oedipus, Oedipus, this cannot be night that surrounds you.

2. The Accident of the Night

Diversity is the material of evolution, since it is from a diversity of beings that natural selection makes its choice. Our genetic endowment differs, as must differ our innate capacity to challenge or adapt to our environment. And no step in the development of life has so contributed to diversity as that reproduction which comes as a consequence of the sexual embrace. The subject of inequality may be unfashionable. But so then was the subject of sexuality itself in Victorian times. If we have today transferred our taboos of discussion from sex to its consequences, then perhaps the progression has come naturally to the puritan mind.

The lower the organism, the less is the distinction of individual identity and the slower the rate of evolutionary advance. As we turn our exploratory steps back through the subdued passages of time, pressing through curtain after fading curtain into the shadowy chambers of life's early mansions, we come to those once-living beings who existed before the coming of sex. The earliest we know left their fossil souvenirs in southern Africa's rocks just over three billion years ago. Life of a sort must have had its beginnings not too long after the formation of the earth itself. Our earliest ancestors have comparable descendants in such durable beings as the amocba who discovered a way of life so successful that they will be with us always.

Diversity developed in those times that we call Pre-Cambrian, but it was not as we know it in higher organisms. Reproduction took place through division, so that mother and daughter cells contained precisely the same genetic material. A clone would develop—an aggregation of single-celled creatures or perhaps, as in algae, a chain of connected cells—and since all members were descended by division from a common ancestral cell, so all were constructed from the same genetic stuff. Change could take place through the random, infrequent action of mutation when some accidental outward force replaced an old gene with one of novel yet workable order. And diversity could come about when clones thus equipped with novel capacities could adapt to novel environments. Perhaps other means of change which we do not quite understand came about, and so other paths to diversity. We may be sure only that change on the whole was slow, and diversity slight.

Our knowledge of Pre-Cambrian times—the ages before the coming of hard-bodied creatures who could leave clean fossil records for the future—has been growing rapidly in recent years but is still of small order. We cannot be sure, for example, just when the first glimmerings of sexual reproduction made their entry into our evolving way. It may be longer ago than we know. But by half a billion years ago sexual reproduction had replaced cellular division as the fashionable means of begetting offspring. Many a conservative organism, while tentatively experimenting with the new way, still clung to the old. But the union of two individuals in contrast to the splitting of one brought mathematical possibilities of diversity to the gamble of survival so decisive as to sweep the living world in behalf of sex.

The American geneticist Sewall Wright showed that an unfavorable change of environment would destroy an entire clone of identical members; but its effect on a population of varying individuals would be to destroy only those least capable of adaptation, while the remainder survived. Species of protozoa and rotifers have been reported that cling to cellular division so long as conditions are favorable, but turn to sexual reproduction when times become hard. The Pre-Cambrian world of backwater slimes, therefore, was one in which offspring tended to be identical and of equal merit. But for the last half-billion years of sexual ascendancy our young have been various, and of unequal potential. And this world of diversity is the world we know.

Such a world was the object of Charles Darwin's observations, as it was of his contemporary Alfred Russel Wallace. How total is the variation of individuals within a species might receive adequate testimony today from the Federal Bureau of Investigation, since no two human beings, not even identical twins, have the same fingerprints. Neither Darwin nor Wallace needed the FBI to inform them, however, concerning a truth observable by any naturalist. But it was Darwin who first suffered the inspiration that was to lead to *Origin of Species*. He was aware of Thomas Malthus' famous work, *An Essay on the Principle of Population*, in which the proposition was advanced that since the number of young in any population must proceed far faster than available food supply, then environmental resource must place a constant check on numbers. As early as 1839, shortly after his return from the cruise of H.M.S. *Beagle*, Darwin put two and two together: varying individuals encountering limited food supply must compete, the most successful and best adapted to their environment thus leaving more offspring for the next generation. This of course was natural selection, among the farther leaps in the history of human thought.

Darwin was not only a perfectionist; he was a procrastinator. To the exasperation of his friends, *Origin of Species* lay still unfinished nineteen years later. Wallace meanwhile was working in the Far East. And in February 1858, bedded by fever in the Moluccas, he seems to have had nothing better to do than think. He too had read Malthus years before, but he was unaware of Darwin's long preoccupation with natural selection. Now, pondering the question of differential mortality, it came to him too that those animals best suited by chance to meet the demands of an environment would be those to survive.

"Then it suddenly flashed upon me that this self-acting process would necessarily *improve the race,* because in every generation the inferior would inevitably be killed off and the superior would remain—that is, the *fittest would survive.*"

On the following three evenings Wallace wrote his essay and with a flourish of excitement sent it off to Darwin in England. Transportation was slow in those days. Darwin received it on June 18, and the sky promptly fell down. For years intimate friends had warned him that just such a catastrophe might take place. Now the same friends, including the greatest geologist of the time, Sir Charles Lyell, came to the rescue. They insisted that one man's devoted labors of almost twenty years must not be destroyed by the priority of another man's flash of inspiration. Darwin must as rapidly as possible draw extracts from his manuscript and present his view.

On the historic if neglected night of July 1, 1858, both papers were presented to a meeting of the Linnaean Society in London. Thomas Bell, president of the society, was in the chair, and the papers, according to custom, were read by the secretary. Darwin was at his home in the country, Wallace was still in the Far East. No discussion followed the reading. For some the subject may have been too new to grasp, for others too traumatic. The president seems to have suffered no shock, for he could later report that "the year has not, indeed, been marked by any striking discoveries which at once revolutionize, so to speak, the department of science on which they bear." But rumors of impending crisis spread through scientific and religious circles.

Wallace withdrew all claims to scientific priority, and almost half a century later recalled, "The one great result which I claim for my paper of 1858 is that it compelled Darwin to write and publish his *Origin of Species.*" This it did. The following year, on November 24, 1859, the first edition was published and sold out in a single day. And the fires of controversy, ignited by Wallace, have not yet been extinguished.

Natural selection in time took its proper place not only in

biology but in any study of life's processes. It offended many in Darwin's day, as its implications offend many in ours. Yet its prime offense is remarkable. Darwin looked to Malthus when he wrote that "the amount of food for each species must, on an average, be constant whereas the increase of all organisms tends to be geometrical." Few objected to the harsh rule of starvation. In our contemporary concern with the population explosion we witness few doubts cast on the proposition that human numbers will expand until meeting the limitation of food supply.

That Malthus was almost surely wrong will concern us heavily in a later chapter. It was not, however, the role of starvation but the role of accident that repelled our thoughts. It offended the religious convictions of the nineteenth century as it offends the social convictions of the twentieth. Perhaps something in our nature finds unendurable the thought that we are accidents of random creation, not portions of some grand design. And while natural selection is in itself a design of magnificent order, perhaps a cowardice within us shrinks from the chore of eternally creating order from eternally re-created disorder. It is all too much.

Yet the random and inevitable variation of beings underlies the theory of evolution as hidden mountains and abysses underlie the seas. Darwin could not explain diversity; he could only observe it. And since he believed that the variations of parents should blend in their offspring, how variation persists became the final puzzlement and frustration of his career. He could not know that only six years after the publication of *Origin of Species* an Augustinian monk, Gregor Mendel, experimenting with common peas which he grew in his monastery garden in Austria, had presented his conclusions to the local Brünn Society for the Study of Natural Science. The Society was polite enough about it and printed his report in their obscure *Proceedings*. But they probably regarded the monk as mad. When two years later he was appointed abbot of his monastery,

Mendel, like a stray scientific cat, vanished from the world's back doorstep as unnoticed as when he had arrived.

While Mendel was precise enough in his description of genes —he called them elements—and of their working in pairs received each from a parent, and of the continued shuffling and reshuffling of these unblending determinants through inheritance, I doubt very much that when he died he knew quite what he had discovered. But the mystery of conception's accident awaited only its resolute detectives. In 1900, as if to outdo the coincidence of 1858, no less than three continental scientists came upon the forgotten paper at once. With the rebirth of Mendel came the birth of genetics. And the accouchement occurred none too soon.

By the end of the nineteenth century the theory of evolution had fallen on evil days. Too many sociologists, following Herbert Spencer's early intimations, transported Darwin's struggle for existence into the human arena, there, in the name of social

44

Darwinism, to consecrate as natural law the more brutal consequences of human might. And too many zoologists, unable to close the gaps in Darwin's thinking, had turned their exclusive attention to the endless chores of species classification. Sir Julian Huxley has described this period with the magical line: "Intoxicated with such earlier successes of evolutionary phylogeny, they proceeded (like some Forestry Commission of science) to plant wildernesses of family trees over the beauty-spots of biology."

With the resurrection of Mendel, life flowed in biology as never before. The theory of evolution began its long move from the status of a persuasive speculation toward the structure of a modern science. Zoologists turned from classifying bird skins in unventilated museums to planting primroses, a healthier devotion. Experiments proliferated. In America, Thomas Hunt Morgan established the cult of the fruit fly, an unattractive beast but worthy of a geneticist's veneration, since it grows to reproductive maturity in ten days. With every experiment, however, complexities developed, contradictions appeared. Then about 1930 the three wise men of population genetics—Fisher, Haldane, and Wright—through mathematical complexity brought mathematical order to the problems of heredity. And population genetics—the hereditary potential of an interbreeding group—provided the solid foundation for modern evolutionary theory.

It has been a long road from a monastery garden to the still later demonstration of the double helix of the DNA molecule. Yet as Charles Darwin, despite the dubious validity of Malthus and his own inability to explain variation, had nonetheless been right about natural selection, so Gregor Mendel out of unaccountable luck had chosen characters in his garden peas that could reasonably demonstrate his laws. Had he chosen others of more complex genetic determination—a character produced by several genes acting in concert, or several characters influenced by a single gene—his calculations would have failed, his experiments would have been abandoned, and we should have had no Mendel to resurrect. Yet despite the simplicities of its one-gene-one-consequence origins (what Huxley with his gift for phrases has called "billiard-ball genetics"), and despite the complexities that later calculations would contribute, Mendelian law explained variation in living beings, and we need look no deeper.

The principle is that of recombination. In the action of sexual conception, two parents with different combinations of paired genes contribute half of the resources of each to what becomes a new genetic proposal. The fertilized egg, this randomly determined recombination of parental possibilities, is the accident of the night.

Theodosius Dobzhansky, an inheritor of Morgan's fruit flies and today, philosophically, the most equalitarian of geneticists, has described the theoretical Mendelian consequences this way: If the parents have 5 pairs of genes, then there are 32 possible recombinations. If they have 20 pairs, then the possibilities amount to 1,048,576. If they have 32 pairs, then over two billion new opportunities confront the offspring. Yet the simplest of animals has genes in the hundreds, and the human being has far over ten thousand. We have no comprehensible mathematics to describe the chances against recombination producing two identical human beings.

We are speaking of theoretical possibilities according to Mendelian law. There are practical limitations. Genes tend to recombine in groups, reducing the chances. Some combinations are so far-fetched that fertilization will not even take place. The British philosopher of science, L. L. Whyte, has described this as internal selection, a process in which the genes, like members of a conservative club, blackball the entry of a fellow too strange. And there is a broad restriction brought about by evolution itself to reduce the waste of accident by limiting the genetic disparity of parents. We shall come back to this later.

Granting all the reductions of theoretical possibility, we must remember that we are reducing odds beyond comprehension to odds still lying beyond comprehension. Let us consider brothers and sisters, the offspring of a single pair of adults. We may, of course, find identical twins, conceived by the union of a single egg with a single sperm. Otherwise, the chance is one in a trillion that any two siblings will be genetically alike.

The accident of the night, which determines in such large part what you or I will be, prohibits identicality. Every month a new egg slips into the female womb, and no two eggs are alike. At the climax of the sexual embrace the male ejects sperm beyond counting into the female genital tract. No two are alike, and in the first trial of natural selection there is no determination as to which sperm will have the superior luck or vigor to win the race to the egg and fertilize it. The magnitude of the competition is such that although only one human being will be the normal consequence of the achievement, still a single teaspoonful of male sperm would be sufficient to father every member of *Homo sapiens* alive today, and all would be different.

Every being conceived by sexual recombination is a genetic accident. Every individual being is thus a pioneer, a biological adventure. No one quite like you can exist in your species. Common heredity may provide a common disposition among contemporaries, or a limited likeness between ancestor and descendant. But the strategy of sex denies the prison of identicality. If you were not created equal, you were yet created free.

2

Henry Allen Moe was for decades director of America's Guggenheim Foundation. I met him for the first time over thirty years ago, when I was a proud possessor of a Fellowship, and I thought that I had never met so gentle, so perceptive, so sensitive a man. A time came when Moe retired to become president of the American Philosophical Society. And a night came, in 1965, when he rose to address Washington's Cosmos Club in response to its award for his contribution to science, literature, and the arts in America. The address might have been delivered from history's most thundering barricade. Few Americans but this gentle, sensitive, perceptive man could have had the courage to pronounce such heresies. In part he said:

> In his *History of English Law*, F. W. Maitland, the most lucid and far-sighted of English legal historians, quotes a text from St. Paul: "It is better to marry than burn." And then Maitland dares to comment, on a text from Scripture, mind you: "Few texts have done more harm than this."
>
> Similarly, from the scriptures of the United States, I shall quote Thomas Jefferson's "All men are created equal" and shall dare to comment that few texts have required more explanation than this. . . . There is no doubt that, at the time it was said, the saying that "All men are created equal" needed to be said, and especially that it needed to be said in the context of the Declaration of Independence. Nevertheless, I shall dare to say flatly now that few unqualified statements have done more harm than this. . . . To say that all men are created equal is—as everybody knows and nobody doubts nowadays, but nobody says—the apotheosis of error.

In his published address Moe points out in a footnote that Jefferson's phrase, which was to become the holy unity of the American secular religion, was made in 1776; and that only thirteen years later, in the Declaration of the Rights of Man, even French revolutionary thought had backed away sufficiently to state: "Men are born equal and remain free and equal in rights."

The French declaration, while false in its premise, left room for maneuver. Jefferson's phrase, presented as a self-evident truth, was false and left none. And so for almost two centuries American thought, with increasing agony and distortion, has been nailed to a cross of revolutionary propaganda, a passing political slogan which its sophisticated author would have been the last to take seriously. Fundamentalist we may no longer be in our religious contemplations. Yet contemporary social theorists can yield nothing to mumbling, illiterate, forgotten multitudes in their cringing devotions to antique screeds.

Sexual recombination imposes diversity on living beings. Evaluated by environment, that diversity becomes inequality. I have described each differing recombination as a genetic proposal. It is the offering of a new, unique being to the test of development in a particular environment. This being, this fertilized egg, this genotype as yet unaffected by dusty skies or the problem of serpents, meets its first tests in the world of the womb. Here it divides and re-divides according to the coded instructions of its genetic material. Fetal life may reveal error in the original instructions; the proposal will be withdrawn and the being aborted. But granted that development is true, and that what once was a genetic proposal has survived life's first sortings to become through birth an independent being, then already we confront superiority. The live are superior to the dead, the born to the stillborn.

Already, too, we confront the phenotype. It is a geneticist's term of salient importance. The genotype is that most independent of beings, the accidental egg, owing allegiance to none but the genetic memories that have combined to create it. The phenotype is the egg plus experience. The womb itself may have failed to protect the developing fetus from experience: undernourishment, drugs, disease, accident, stress or perhaps alcoholism in the mother. Even in the womb, environment will have imposed certain alterations on genetic destiny. The newborn infant is already a phenotype, a genotype modified by environment. And as a world of diverse experience opens up before the newly independent being, so will the interactions of genetic determination and environmental persuasion bring new courses, new colors to the phenomenon of individual existence.

The accident of conception commands that no two beings be genetically identical; equally so, the accidents of existence command that no two lives proceed along identical ways. The diversity of our beginnings is compounded by the diversity of our histories. But there is a difference of large order: whereas randomness dictates the accident of our origins, nothing like randomness dominates the paths, however various, that our lives will pursue. Any population of men or deermice, of vervet monkeys or herring gulls, faces in a state of nature approximately the same environment presenting approximately the same demands, the same hazards, the same opportunities. A diversity of beings encounters a singleness of being. Disorder encounters order.

It is this singleness of the playing field that reveals the inequality of the players. To a degree, the environment promotes equality in the elimination of hopeless variants. If you are a young zebra who cannot keep up with your family, then the lion will get you. If you are an ailing caribou, then you belong to the wolf. All survivors have something in common: that they have survived. To a degree, also, environment promotes equality in its shaping of phenotypes. Various though we may be in our beginnings, we learn our lessons from the same master. Yet the shaping of muscles and manners and mind to uniform environmental demands can go only so far. Experience may fulfill or reduce the potentialities of the accidental egg, but it cannot create what was never there. And we cannot leave to our descendants the mind or muscle we have acquired in the exercises of survival; every egg comes fresh.

That acquired characteristics may be inherited, as most of us know, was the interpretation given to evolution by the French naturalist Lamarck. In Darwin's inability to explain variation he was frequently tempted to embrace the idea, yet, fortunately, for the most part resisted it. Lamarckism has been disproved in the laboratory over and over again. Generations of mice have surrendered their tails, only to grow new ones. It has been acidly observed that the sacrifice of mouse tails has been unnecessary since for thousands of years the Jewish people have been cutting off foreskins without noticeable effect on their male offspring. Even so, it is today a portion of the naïveté of Marxism that a new kind of social system will produce, permanently, a new kind of man. It is Lamarckism, and Lysenko was his Soviet prophet, installed with halo and power by Stalin. The power banished Western genetics from Soviet science, as it banished N. I. Vavilov, among the world's greatest geneticists, to a Siberian labor camp, where he died in 1943. Lysenko may today have vanished from Marxist power; yet his memory lingers as a portion of Soviet agricultural inadequacy. Vavilov may be dead of arguable causes; his memory fails to linger in those minds who in adolescent reverie equate Marxism with progressivism, and competition of men and ideas with the political counterrevolution.

A science as mathematically remote as genetics may become the wheel of political fortune. Yet one must admit that, were it not for a force of pervasive disposition, temporal power in all brutality might by environmental harshness reduce even the newborn, in terror, to phenotypic mediocrity. To safety, to conformity, we might flee days after birth. But it is an evolutionary impossibility described by Konrad Lorenz in *On Aggression*. Whatever environment's iron hand, as Lorenz sees it, every being will challenge such rule, will seek to achieve non-identicality and to fulfill its diverse genetic potential through an aggressiveness inborn. The drive to fulfill oneself, to perfect if possible the genetic potential of one's unique endowment, is itself coded in our genetic instructions.

We might say, in the context of this inquiry, that aggression is the natural guarantee placed in species that disorder—the richness of diversity generated by sexual recombination—will not fail of development. The London psychoanalyst Anthony Storr, brilliantly expanding the Lorenz thesis in his *Human Aggression*, writes of children:

> There is the need to cling to the mother, to be sure of her affection and support. But there is also a drive to explore and master the environment, to act independently. . . . One important function of the aggressive drive is to ensure that the individual members of a species can become suffi-

ciently independent to fend for themselves, and thus in their time to become capable of protecting and supporting the young which they beget.

And Storr cites the American analyst Clara Thompson in her conclusions on the same theme:

> Aggression is not necessarily destructive at all. It springs from an innate tendency to grow and master life which seems to be characteristic of all living matter. Only when this life force is obstructed in its development do ingredients of anger, rage, or hate become connected with it.

Aggressiveness, we may then suggest, brings to life as its partners rebellion and competition. We need not think in terms of the Darwinian "struggle for existence," or the nineteenth century's "survival of the fittest." No such lethal stakes are involved, and Lorenz rightly points to the rarity of corpses decorating the scene of animal decision. The vindication of the individual, not the annihilation of enemies, is aggression's goal. But such aggression commands competition. I return to Storr:

> The normal disposal of aggression requires opposition. The parent who is too yielding gives the child nothing to come up against, no authority against which to rebel, no justification for the innate urge towards independence. No child can test out his developing strength by swimming in treacle.

The aggression coded in our genes compels the full development of the individual, a natural necessity; and the diversity of individuals all comparably coded compels competition, without which natural selection could not take place. And, finally, the competition among fellow members of a population not only brings to maturity, for worse or better, the genetic fortune of our origins, but through endless sortings evaluates unequals in terms of environmental demand.

"Human evolution is based on injustice," wrote Sir Arthur Keith a quarter of a century ago. I rarely disagree with that most admirable of anthropologists, but Keith was wrong. The evolution of men, like the evolution of meadowlarks, is based on the recognition and adequate sorting of unequals. Injustice occurs when competition is aborted. Injustice occurs when worth fails of recognition, and the unworthy go rewarded. While injustice may be regarded as a principal feature of human life, raised by human ingenuity to altitudes unknown in the world of the animal, it is not evolution. In terms of natural selection, injustice is maladaptive.

I have asserted that equality among sexually reproducing beings is a natural impossibility. I may assert with comparable force that in the history of vertebrate life, which has brought man to ascendancy, equality of opportunity is a natural law. It is opportunity suppressed not just by the totalitarian, but by the equalitarian as well.

Two major experiments have enlightened evolution throughout the past half-billion years. The first has been that of the insects and such sea-going cousins as the lobsters, all of whom have soft bodies with their hard parts on the surface. The second has concerned the vertebrates, the anatomical reverse, whose soft bodies surround hard inner backbones. Sometimes we digressed to form hard outer armor, like the turtle and the armadillo. But in the backbone resides our strength, from earliest fish to most recent lion or monkey or bird.

While the vertebrates came to dominate the living world, their triumph is less than clean-cut. According to Simpson, of the million-odd species of successful creatures alive today, only forty thousand are vertebrate. The insects may, at some moment of future disaster, survive us all. But the experiment has tested behavioral as well as anatomical divergence. The insect, by and large, has depended for behavioral guidance on programmed instinct, the vertebrate on programmed learning. And the final returns on the success of intelligence have not yet been counted.

The insect, of course, is far from being entirely dependent on those behavioral instructions coded into its genetic beginnings. The Dutch ethologist G. P. Baerends once demonstrated the awesome capacities of the digger wasp to learn and remember. The female normally digs a hole, kills or paralyzes caterpillars, buries them and lays eggs on them. She then proceeds to dig second and third holes, where she repeats the process. In the meanwhile the eggs in the first hole have hatched and the young have been busy consuming the caterpillars. She opens each hole, restocks it.

One might interpret the mother digger-wasp's actions, at this point, as wholly instinctive. But Baerends found that she opened all three holes the first thing in the morning before proceeding on her caterpillar-catching rounds. And so before she appeared he robbed from one hole, added to another. What seemed to be demanded normally was six or seven caterpillars. And so to those holes he had robbed she brought more, to those to which he had made a contribution she brought less. Not only was she able to discriminate between the numbers demanded by her early-morning appraisal, but through as much as fifteen hours of hunting she remembered the differing demands of her differing nests.

Niko Tinbergen, Oxford's pioneer ethologist, has commented that, high though this order of learning may be, it is localized and applies only to a single, instinctive activity. We cannot say that the digger wasp has learned to count. What she has done is to complete from experience a remarkably demanding instinctive program. And this specialization of capacity may be the basis for caste in social insects.

As vertebrates have developed our societies, so insects have theirs. Since the turn of the century entomologists have revealed the wonders of social organization in the ant, the termite, the bee. Most famous of all, perhaps, has been Karl von Frisch's investigations of honeybee communication through dances. And in all such societies the member is born to his role, or succession of roles. The presence of castes has made known to us just how ancient was the evolution of the earliest societies. Recently in New Jersey a piece of amber was found and its age established as 100 million years. Within the amber, sealed as if for exhibition on Judgment Day, was a worker ant.

The evolution of living systems governed by programmed instincts finds selective advantage for social life in specialized classes of beings, not in individuals. But the evolution of living systems based on programmed learning finds a different selective advantage in the random excellence of individuals, not the destined appointments of caste. In vertebrate societies one finds two roles biologically determined, those of male and female. Otherwise the individual is on his or her own. So rare are the exceptions that they become worthy of consideration.

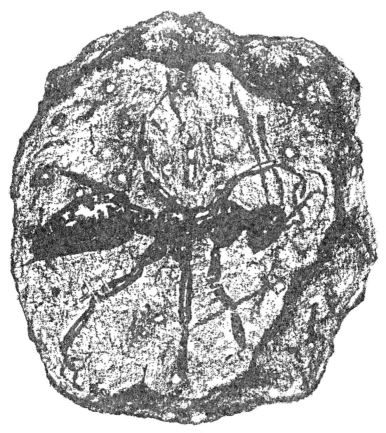

The ruff is a bird whose antics I described in *The Territorial Imperative*. In the spring aspiring males develop in all gorgeousness a shield of feathers about the neck, from which they derive their name, of such various colors that no two birds look the same. They then adjourn to what are known along the Dutch coast as hilling grounds. Here, each on his tiny hill, they strut, gyrate, display to one another, and fight. Later the hens, known as reeves, will come for copulation. But more recent observations have revealed subtleties of behavior previously unknown.

Among the males are two classes, the territorial males, true gladiators, and satellite males. The dominant, territorial class has a ruff of lighter shades of color, linked with a highly aggressive disposition. The satellite has a darker ruff, avoids conflict, and submissively stays in the vicinity of the master in a parasitic role. Tolerated by the territorial male, the satellite is occasionally permitted a copulation or two. The conclusion of the observers has been that, like blue eyes and brown, the classes are under genetic control. The linkage of coloring and behavior is brought about by polymorphism, the geneticist's term for determination of several characters by a single pair of genes or group of genes acting together. Caste in the ruff is not directly inherited. It comes about through the usual random sexual recombination, but in every generation according to a regular mathematical proportion.

If the conclusion is correct, then the ruff exhibits the only genetically determined class system within a sex that I have ever encountered in the vertebrate world. There are examples of social determination, however, reminiscent of human affairs. These occur in related species of monkey, the Japanese and the rhesus, and may occur in more.

The Japanese monkey is the social genius of the primate family, and we shall turn to him for many later illuminations. Studied constantly by a group of Japanese scientists for many years, the monkey lives in some of the largest organized societies achieved by any primate, human or subhuman, before man began to gather in towns. In these colonies, males sort themselves out in rank orders of dominance, a state of affairs common in monkeys. Uncommon, however, is a similar order among the females. And observation indicates that unless the colony becomes too large the son of a high-ranking female has a far better than normal chance of achieving high rank among the males. Kinji Imanishi, reporting on the studies, concludes that a son who has learned to get along successfully with a highly dominant mother will have the greater success cooperating with equally dominant males.

The related rhesus monkey of India is the most studied of primates in laboratories. Thirty years ago C. R. Carpenter, the pioneer student of primates in the wild, transported a large number of animals from India to tiny Santiago Island, just off the Puerto Rico coast. The colony he established still thrives, still is studied by various observers. And here has recently been recorded behavior comparable to what Imanishi watched in Japan.

The rhesus too lives in quite large organized bands within which females have orders of dominance in relation to each other. And in most bands maturing sons of dominant females succeed rapidly in outranking many elders in the male hierarchies. But the observer Carl Koford reports an odd outcome perhaps reflecting on such arrangements as unnatural. Two princelings, having achieved high rank, shortly vanished from their hierarchies and their bands to become solitaries, later to join other bands in positions of low order. "High birth permits rapid advance in the social hierarchy, apparently, but it does not insure succession to leadership."

It is a subject about which we shall know more one day. At present, however, these are the only two vertebrate species ever to come to my attention in which social discrimination takes place. And never did exceptions so prove a rule. Equality of opportunity is a vertebrate law. The castes of sex are universal, creating the simplest expression of division of labor. But within each caste the sorting of unequals takes place in a field without fences.

Animal justice was perhaps the first natural law that civilized man began systematically to violate. Advantages of birth offer no guarantee of genetic superiority. Restrictions of caste, of class, of occupation, of poverty distort or suppress the phenotypic flowering of genetic endowment in the maturing individual. But the accident of the night, in all its rich, random resource, became in man socially aborted. There have been revolutions, it is true. But human history has far more frequently witnessed the decline of empires, the vanishment of kingdoms, the disappearance of peoples genetically exhausted through order's injustice.

We shall turn to the inequalities of populations in a moment. But first we should reflect that animal justice reflects the fairness, the freeness, and the openness of the competition of unequals. Did inequality not exist, justice would have no function. And perhaps that is why the equalitarian ideal in human life, denying the nature of man, moves so easily into the tyranny of thought and power.

3

What in genetics looms as the Class of 1930—Wright, Fisher, and Haldane—confirmed with their mathematics the general principles of Mendelian heredity. But they did more: they established the science of population genetics. Following them came a generation of new explorers of the gene—Dobzhansky, Ernst Caspari, C. D. Darlington, Kenneth Mather, C. H. Waddington, J. M. Thoday, many others—leading on to the molecular explosion of James D. Watson and Francis Crick. A luminous volume for the lay reader is *The Language of Life*, by George and Muriel Beadle. Dr. Beadle is a Nobel Prize winner in genetics, and former president of the University of Chicago. And the first sentence of their foreword reads:

In writing about genetics for non-scientists, the gap we have to bridge is not the one that is alleged to separate C. P. Snow's much-publicized two cultures, but the one that lies between people who received their formal schooling before the mid-fifties, and those whose instruction in science has occurred since then.

Respected figures in academic life, when speaking of the natural sciences, frequently sound as if they had received no schooling at all. In all compassion we must turn from such a conclusion. It is just that they went to school too soon.

Population genetics turned biology inside out. Harvard's Ernst Mayr has written that "the replacement of typological thinking by population thinking is perhaps the greatest conceptual revolution that has taken place in biology." But the revolution, like a nuclear explosion, has produced a fall-out far beyond biology's immediate neighborhood. It has affected economics, and its onetime devotion to the typical, economic man; it must affect behaviorist psychology and its reluctance to accept human variation; it must affect this chapter and our preoccupation with the accident of the night.

Population thinking denies uniformity and looks to the *range* of diverse individuals within a group. The range, not the average, is the reality. Yet in his classic *Mankind Evolving* Dobzhansky writes: "The assumption of the psychic unity, or uniformity, of mankind is probably pivotal in the working philosophy of a majority of anthropologists, psychologists, sociologists, and not a few biologists." Such a uniformity is as impossible in mankind as is identicality in the members of a Boy Scout troop. Just as population thinking accepts range as reality, it dismisses as nonexistent the "average man," a being whom no one has ever met anyway.

Population genetics was a wild child of mathematics conceived in a forest of impenetrable equations. It has emerged from that forest to invade every street corner of thought as a philosophy that speaks of probability, not certainty, of horizons, not pigeonholes, that acknowledges change and rejects the immutable and accepts disorder as the foundation of order. It is a way of thinking obviously more sophisticated than our primitive, childlike devotion to the type. We may not be equal to it. Yet population thinking is indeed an imitation of nature, a philosophy in accordance not in conflict with natural processes. And that is why it is revolutionary. The supreme contribution of twentieth-century biology may only in passing concern what men know; in final assessment it concerns how men think.

The accidental consequence of sexual recombination has been a problem to nature and not just to naturalists. So uncountable are the chances of diversity, as we have seen, that sexual union, without restriction, could have launched living beings on a voyage into chaos. And so the first of the restrictions, the isolation of species, took on the burden of containing accident within reasonable bounds. Sexual union between members of separate species normally but not always results in offspring that are infertile or of declining fertility. Through "reproductive isolating mechanisms," however, the nearly total likelihood is that such sexual union will not take place.

These isolating mechanisms, ensuring that girl dog will shun boy cat, are as varied as species themselves. Some are anatomical, such as that of the red-faced stump-tailed bear macaque, *Macaca arctoides*, whose penis is as improbable as his name. The bear macaque, the crab-eating macaque, and the far more numerous rhesus monkey are three closely related species of the same genus frequently living in the same area. And the species problem of the bear macaque, it seems, has been how not to get hybridized out of existence by his crowd of relatives. The solution has been anatomical, and one of rare device.

All primates, including man, have in the male a penis of much the same appearance, including the blunt, helmet-shaped glans. For a long time, however, observers of such unimaginative male decor have wondered at the iconoclastic penis of the bear macaque. Its glans penis is long, slim, tapering like an aristocratic finger. Moreover, the penal bone is at least twice as long as in other macaques. (In all human regret, it should be noted that man is one of the very few primates unaccountably denied this anatomical luxury.) The question of why the male bear macaque should have such a penis has been answered only recently by intimate investigation of the female bear macaque. Across her vagina is an obstruction effectually blocking the penis of any but her slenderly built fellows. It is a chastity belt of selective advantage to bear macaques.

Konrad Lorenz has described as lock-and-key such sexually isolating devices, but they are usually less literal in design. In northern Kenya, for example, two species of zebra sometimes form mixed herds. The common zebra of the East African grasslands, with its broad stripes, is the one familiar to visitors and has the species name of *burchelli*. Another species, *grevyi*, has much narrower stripes, lives in the arid lands to the north and sometimes in drought moves south for water. It is then that they mix. Yet no known hybrid has ever been observed, nor attempted copulation between species witnessed. The differing patterns of female rumps apparently fail as sexual releasers.

Behavioral differences may in themselves isolate species. In 1963 Bristol University's John Hurrell Crook published an elegant study of weaverbirds. They are colonial creatures inhabiting a single tree. Each male weaves a globular nest which hangs neatly from a branch, so that the tree seems to bear a crop of most remarkable fruit. The yellow weaverbird, *Ploceus cucullatus*, has a tendency to choose a tree near houses or villages, probably to enhance protection from such predators as the genet cat and various snakes; and so at an East African lodge you will frequently find the flurry and chatter of nearby weaverbirds an entertainment for which you are not billed. But Crook studied weaverbirds in West Africa as well, where near Lagos he found the black weaverbird, *P. nigerrimus*, nesting in mixed colonies with the yellow in tall oil palms.

Differences of color are perhaps enough to keep the species from crossing in such close domestic arrangement. But the sharpness of behavioral difference makes one suspect that color is not enough. Unusual in bird life, the males of both species are polygamous, building several nests before acquiring female tenants, and defending the whole space as a territory. Even in this male defense, behavior differs. The yellow advances on an intruder, lunges, then, at his boundary, sings and finishes with a deep bow. The yellow intruder does the same, advancing on the boundary, singing and ending with a bow. Here, a foot apart on a branch, like the most gallant of old-time swordsmen they exchange the thrusts of their songs, the deprecations of their bows. But occasions arise when the field of honor becomes somewhat confused, and such moments come when the black is the intruder, the yellow the defender. The black has no such song and bow in his behavioral repertory. Instead he does an agitated dance. And so at the boundary we find the yellow weaverbird singing and bowing, and the black weaverbird hopping about in his dance. Yet honor, despite differences, remains accomplished.

A comparable disparity of behavior between females of the two species may account for sexual separation. At time of courtship the yellow male hangs from one of his nests below its entrance, advertising his desirability as mate and landlord by pumping his wings. A female flies through the tree, setting off an explosion in the colony which attracts more females and more violent waves of exhibition. But there is no aggression. If she enters a nest, then the issue is settled; if she passes it, the male does not argue. In the black species, however, sexual excitement is aroused by quite other means. When the female crosses a territory, the male pursues her, kindling excitement in other males, who join the chase like Roman *ragazzi* in pursuit of a blonde. But if the weavers have picked the wrong female, the yellow has only to pop into the nest of a bird who has not chased her. She will be with her own kind.

One last, uneasy example should be added to the range of natural ingenuity concerning reproductive isolation. The sight of snails copulating on a lawn at night is a common observation of suburbanites. And the question might well be raised:

How can two snails of differing species be aware of mistaken identity, particularly in the dark? The answer has been facilitated by a piece of equipment peculiar to the sexual life of the snail. It is called *spiculum amoris,* or the love-dart. It is a slender, blade-like instrument made of the same limy material as the shell. In the act of copulation one plunges his love-dart into the flesh of his partner, precipitating a peak of sexual excitement. But the response may be otherwise. If members of two related snail species, *Cepaea hortensis* and *C. nemoralis* attempt copulation, then problems will arise. *C. nemoralis* has the larger love-dart and thrusts it more powerfully. If the partner is of the weaker species, then the frightening dart will break up the affair.

Love is not so blind. Sexual discrimination has been a product of natural selection, restricting union to those animal groups called species with sufficient genetic material in common to make probable the viability of offspring. But while any two members of a species may in theory mate and breed, in fact normal choice is more limited. Evolution has encouraged a further subdivision, the population, to provide still more ordered diversity.

As our foremost authority on species, Mayr has chosen as his favorite definition that of Sewall Wright: "Species are groups within which all subdivisions interbreed successfully to form intergrading populations wherever they come in contact, but between which there is so little interbreeding that such populations are not found." In other words, Wright accepted as the true evolutionary unit the freely interbreeding population which to a degree interbreeds with adjacent populations until one comes to the species border where no further such populations are found. In his *Animal Species* Mayr, like Wright, looks on the relation of populations, not individuals, as the definition of species. And so we turn to the population.

Simpson has given us the favored definition: "A population is a group of organisms, similar among themselves and dissimilar from other organisms with which they live, descended from a not remote common ancestor, living in the same general region and environment, and with a continuity of many generations through time." The American psychologist John B. Calhoun has introduced another dimension: "By a population we mean any contiguously distributed grouping of a single species which is characterized by both genetic and cultural continuity over several generations." He has added the significant comment, particularly in human terms, that the population is "the major unit in which the lives of individuals find reality." We may think in our species of nations and tribes as such major units.

For the purposes of our discussion, I believe it sufficient to describe a population as a group of beings sharing a generally common environment among whom male and female have a variable but probable opportunity of meeting, mating, and breeding; where that probability drops sharply, there is the limit of the population. Consider two African tribes like the Kikuyu and the Masai living in adjoining regions in East Africa, but separated by language, by tradition, by custom, and by the differing ways of life of farming and pastoral people. A certain interbreeding has taken place as a consequence of the Masai raiding Kikuyu villages for wives. But they are distinct populations physically and psychologically, in each of which the individual finds his major unit of reality in the tribe, not the country called Kenya. The heart of the problem of independent Africa has been that the border of the population rarely corresponds with the border of the politically created nation.

In the human species, language as a rule forms the sharpest barrier between populations, and that is why the line of language forms usually the national boundary, and why, with rare exceptions, political boundaries enclosing varying language groups tend to enclose trouble as well. But even within an interbreeding population of common language, still religion, geography, allegiances of custom or occupation, economic class, even degree of education all tend to reduce to smaller groups any high probability of meeting and mating. And so the smallest unit of truly equivalent chance is known as a deme, or isolate, and Dobzhansky has suggested that among people even today isolates number hundreds, rarely thousands. A study of marriage records in France has revealed the paradox that in the mountains such isolates average about 1,100 while in Paris, where one would expect broad mixture, they are only about 900.

Within such isolates gene flow is complete. Their existence cannot, however, be explained as a human consequence of our cultural diversity, for in animal species the isolate may be of remarkably similar size. Sage grouse in Wyoming, isolated by allegiance to a single strutting ground where all copulation takes place, number about 900. Helmut Buechner's Uganda kob, which I have described elsewhere, are likewise divided by traditional stamping grounds into reproductive groups of a 1,000 or less. George Bartholomew's study of a breeding colony of

Alaska fur seals on St. Paul Island gave us 800 cows and 40 bulls. Two separated wings of Tanzania's Serengeti reserve have each about 900 elephants. Since the fluctuating members of all-male parties in the western group seem familiar with one another, we may assume that it is a reproductive unit.

Wright, Simpson, and many others are agreed that the greater the subdivision of a population into such reproductive isolates, the better is the chance for survival of the entire population. Each isolate, through generations, develops its own pool of genes of similar character in which the accident of unfavorable union is reduced to a minimum, while sufficient numbers retain sufficient diversity. A chance mutation of benevolent order will have the opportunity of spreading to all members of such a restricted group. Drift—a gradual genetic change of no specific cause—will through the generations lend the group a special genetic character. Whatever may be the fate of the isolate, the selective advantage for the whole population is immense. We have seen that the population, despite its mosaic of breeding groups, faces a fairly uniform environment. But as no two individuals can be quite the same, so the gene pools of no two isolates can be identical. With any change of environment— a drying up, an invasion of predators, a change of food supply, a pestilence—chance must surely determine that some groups will be superior to others in meeting the new challenge. While the inferior may be lost, a portion should survive to restock the population with its more adaptable endowment.

Out of the concept of time—out of the image of a population's survival not just in a given space throughout a given generation, but through changing conditions and perhaps an infinity of generations—has come the radical departure of population thinking. The individual may die; in his contribution to the population he becomes immortal. Individuals may vary; in the range of their variation lies the resource of the population. Individuals may fail, or succeed; the sum of their successes means survival of the group. The life of the individual may be inconspicuous; but his genetic contribution to the population may with changing times be of critical value. It is the population, not the individual, that evolves.

The co-adapted gene pool—the whole population's reservoir of genetic possibilities—has been described by Waddington as the most advanced concept yet reached by contemporary evolutionary thought. Natural selection has gradually restricted the pool of genes to those with a fair chance of getting along together. Yet through one device or another—through the diversity of isolates and such hidden assets as the recessive gene—it maintains its ramparts against the unknown future.

Heterozygosity is a forbidding word, but since there exists no synonym, we must use it. Heterozygosity describes that pair of genes in which one is dominant and displays the character to which it contributes, and the other is recessive and displays nothing. You may have brown eyes. If you are homozygous—that is, if both genes are the dominant brown—then you can have only brown-eyed children. But if you are heterozygous, then anything can happen. Kenneth Mather demonstrated a long time ago that homozygosity governing most characters in a population better fits it for present conditions, but the hidden recessives of heterozygosity lend it better chances for the future. Natural selection has more material to work on, and in a time of stress the population, in Waddington's phrase, may be able to pull something new out of the hat.

Several geneticists, and in particular J. M. Thoday, emphasize today the startling change that has come to the old cliché "survival of the fittest." Population philosophy demands that we speak of fitness for today *and* tomorrow. Whether we are selecting a group of organisms, a field of ideas, or a political party, we must judge not only the compact answers for present survival but the inherent potential capacity for change to meet unknown future contingencies. Extinction has been the normal destination for species or populations bearing only one answer, as defeat is the normal lot of the general who has only one plan. Variation: the variant individual who makes little sense in today's climate, but who may save us in tomorrow's; diverse isolates, spreading the risks of total population commitment; the recessive gene, hidden here, hidden there, waiting for new environments to perform the selective alchemy of transmuting dross into shining metals. Population philosophy brings precision to the evaluation of disorder.

The new way of thought brings likewise a yardstick for populations. Since the time of Melville Herskovits and his cultural relativism it has been fashionable in anthropology to regard all cultures and all human populations as equal. The argument was persuasive. Since environments differ and each population presents its own answer to the challenge it faces, one can scarcely discriminate among answers. The most primitive tribe may, indeed, adapt more successfully than the most advanced nation to inescapable environmental demands. Yet cultural relativism rests on the concept of fitness that prevailed before population genetics. It acknowledges a single standard, the adaptability of a people to its present situation. Cultural relativism ignores the capacity of a people and its culture to evolve successfully in the face of change. "The variation within a population constitutes the raw material of evolution," wrote Haldane in 1949. This variability, in a time of rapidly increasing change, may be the population's single greatest asset.

A population must achieve a fair degree of adaptation to its environment if it is to survive in the present. And if fitness for today were the sole criterion, then cultural relativism would be theoretically sound. But adaptation can be too perfect. When selection for conformity has persisted through a sufficient number of generations, all may seem well, yet reduction of variants will have affected the population's gene pool and reduced its prospects of survival tomorrow. Either variation so wild as to render present survival dubious, or conformity so narrow as to endanger the future, becomes the character of a genetically inferior population.

We may discard what Henry Allen Moe described as the apotheosis of error, the doctrine that all men are created equal; but we must likewise discard as fallacious the proposition that all human populations, even in relation to their own environments, have equal potential.

4

An inquiry into human inequality cannot be regarded as concluded if it fails to enter the lethal arena of race. Yet little will come of this most fateful of entrances but the spectacle of the sciences bearing a white flag. Despite all claims, we know almost nothing. The natural sciences may properly regard the individual as a biological entity, a genetically random accident shaped and sorted by environment. The population too is a biological entity, since it possesses a common pool of genes separated from other pools by that borderline where interbreeding seldom occurs. Individuals and populations may therefore be compared in terms of disposition or indisposition to meet the challenges of an environment, either present or future. But a race comes about not by reproductive but geographical isolation. The racial confrontations of modern man could occur but rarely in a state of nature.

A race, or subspecies, is a mosaic of populations which has

followed its own line of evolution through many generations separated from other such mosaics by natural barriers. During the last ice age of the Pleistocene, let us say, a species of European bird frequenting woodlands may have retreated before the advancing ice and tundra to wooded havens far apart where the species could yet survive. Divided for tens of thousands of years, facing different conditions of natural selection, the separated groups developed distinctive colors of camouflage or display, perhaps certain distinctive behaviors. Thus they became subspecies, identifiably different, yet not so disparate as to prevent interbreeding should contact be resumed. Then the ice and tundra retreated, the woodlands restored comfort to valleys and plains long forbidding. The birds returned. Contact of populations was restored and gene exchange along the margins brought hybridization to border groups. After a few thousand years one might still find a degree of racial purity in racial heartlands, but hybridized populations would link genetically the species as a whole. One would find no clear borderline.

Human races evolved in geographical separation throughout a time before wheels and boats, before sophisticated weapons and tools, before domesticated animals and crops, when barriers like water and mountains and distance itself were formidable indeed. But then, just as the withdrawal of glaciers brought the races of birds together in regrown woodlands, so cultural advances reduced natural barriers and encouraged the mixing of peoples. Probably hybrid populations had always existed along slim lines of racial contact. Now they spread.

If we briefly consider black Africa—that portion of the continent lying below the Sahara—we may grasp the complexity of even this most homogeneous of modern races. Migration played a part. Until about the time of Christ, the black race—what we regard as a human subspecies—was confined to West Africa, excepting only a branch that seems to have spread along the Sahara rim to the Upper Nile. All these peoples farmed dryland crops, chiefly millet and sorghum. Then quite suddenly a change came. Malayan and Indonesian sea trade was beginning to arrive at the East African coast, bringing such plants as bananas and yams. They were crops of superior nutrition adapted to wetter lands. A population explosion took place in West Africa, together with a freedom to move into the damper forest areas. And so, about two thousand years ago, the migrations east and south began to populate all of the sub-Saharan continent which we today call "black Africa."

And hybridization began. We know little about the pre-Negro populations of Africa. The yellow-skinned Bushman of the Kalahari Desert is a relic people, and he seems at one time to have occupied lands from East Africa to the Cape. There were probably other such primitive hunters. The migrating blacks mixed here with this people, there with that, to form what today is known as the Bantu. It is incorrect to speak of any but the West African or the tribes of the Upper Nile as Negro. Such celebrated tribes as the Zulu and Xhosa in South Africa and the Kikuyu and Baganda in East Africa are hybrid peoples with languages and physical characteristics distinct from the pure parent race. If we of the west are unfamiliar with them, it is only because our slave trade drew almost exclusively from the Negro of West Africa. Arab slavers transported the Bantu to the eastern world.

Thus the first division of the race we call black is between Negro and Bantu, and the genetic origins of the Bantu peoples may, for all we know, be as varied as they are obscure. But another migration, this from the northeast, created entirely new groups. Hamitic peoples, Caucasian in origin, had crossed from southern Arabia to spread through Somalia and Ethiopia and to hybridize in varying degrees at varying racial borders. Many were cattle people, and five or so centuries ago began drives of conquest to the south. Such are the Masai, the Watutsi, perhaps even the Herero in faraway South West Africa. Tall, proud, warriors by tradition, they formed island populations in the Bantu sea. Perhaps they derive their extraordinary stature from an early mixture with the tall, pure Negroes of the Upper Nile. Whatever their origins, no lines of racial animosity can rival the contempt with which these peoples, themselves a product of hybridization, view their longer-settled Bantu neighbors.

Yet all is black Africa. What we regard as a race is an intricate mosaic of some thousands of tribes, each an interbreeding unit with a common history but isolated from its neighbors by language, by widely varying customs and traditions, and, as a rule, by hostility. Contemplating such African diversity, one is tempted to sympathize with the University of Michigan's Frank Livingstone, who entitled a paper "On the Non-existence of Human Races."

The population with its co-adapted gene pool has measurable reality for the scientist. The overwhelming environmental change which independence has introduced provides overwhelming disproof for the acceptances of cultural relativism. Some populations, such as the Kikuyu in Kenya and the Ibo in Nigeria, have contained superb potentiality for change. They were fit for tomorrow. (The Ibo, indeed, proved so fit that no recourse was left to their inferior neighbors other than to slaughter them.) But some populations have so far demonstrated little or no such potentiality. What can the natural scientist say of such a natural mosaic of peoples?

Tempted though we may be to dismiss races as non-existent, still we cannot. Had Livingstone come within even shouting distance of the truth, then puzzled indeed must be the spectator at the Olympic Games as he watches the representatives of no more than 7 percent of the human species staggering off the field of honor scarcely able to lift the weight of their medals. The observer must conclude that if race and genetic racial differences do not exist, then awards must in some most devious fashion have been rigged by the United Nations Assembly.

The athletic success of a very small portion of the human species (even including the white-hybridized blacks of the Americas, there are only a few more members of the race than there are citizens of the Soviet Union, and less than half as many as the inhabitants of India) recalls Fisher's comment that natural selection is a device for generating a high degree of improbability. And improbable is the word for the black. Despite all hybridization, all cultural disparity, all environmental divergence, such common traits as superb teeth and the capacity to run forever or jump over the moon or knock a baseball from San Francisco to Los Angeles must find an explanation in some dominant genetic complex inherited from common West African ancestors.

The problem of race is not that it is a fiction, but that it has never been invaded. Not even in ethology has the varying behavior of animal races received proper attention. A notable if largely forgotten exception was an experiment in 1950 by a man named Van T. Harris who was taking his doctorate at that citadel of small animals, the University of Michigan's zoology department. Harris later dropped out of the sciences, and I do not believe that his dissertation was even published. Two years later, however, his work was cited in detail by John Calhoun, and at a still later date was re-investigated. And since it is at once a remarkable story, a confirmed experiment, and a rare excursion on the part of the natural sciences into racial behavior, I suspect the study may be around for quite a long time, still puzzling the experts, still furrowing brows.

Peromyscus maniculatus, the deermouse, is a species adapted to so many differing habitats that in North America it is divided into sixty-six subspecies, somewhat outdoing *Homo sapiens* himself. Two of the most common races are the prairie deermouse, *Peromyscus maniculatus bairdii*, and the woodland race, *Peromyscus maniculatus gracilis*. The prairie form has a short tail and short ears, and refuses to enter wooded areas even though a grassy carpet lies beneath the trees. *Gracilis* has a long tail and long ears, and is as intolerantly devoted to his woodlands. Yet no physiological differences seem to dictate their environmental preferences. They interbreed freely in the laboratory, but almost never in a state of nature, where geographical attachment divides them as would a sea.

No one would suspect the deermouse of possessing much of a psychology, but Harris put it to a strange test. He had been gathering up specimens of both races through the late 1940's, breeding young ones in his laboratory, where they gained no experience of grassland or woods. Then he arranged two experimental rooms with a tunnel for free connection. In one he spread thin strips of paper crudely imitating grass; in the other he placed upright branches. Now he introduced adults which he had caught in the wild; and it was startling enough that they would find in his imitation woodland a familiar environment, in his clipped-up paper strips a semblance of the prairie, and so they would sort themselves out. What was to haunt scientists, however, was that generations born in the laboratory would make the same choice, *gracilis* to the branches, *bairdii* to the papered floor, though they had never experienced woodland or field.

Calhoun puzzled over the aesthetic problem that deermice could accept a representation of an environment; but later students at Michigan seem to have wondered if Harris had not made some mistake. And so in 1962 another man, Stanley Wecker, set out to discriminate between what was learned and what innate in the racial psychology of deermice. He found an oak-hickory woodlot sharply edging an open field, and here built a pen ten feet wide and a hundred feet long, half in the woodlot, half in the field. He divided it into ten sections, each barrier with a little stile over which his deermice could cross freely, together with an electrical recording device to keep accurate track of comings and goings. For subjects he had not only wild-caught prairie deermice and their laboratory-born young, but descendants of Van Harris' prairie race, laboratory-raised for as much as twenty generations.

The experiment confirmed that of Harris years before. The wild-caught favored the pens away from the woodlot, as did the younger generation raised in the laboratory. But it was the experiment with the descendants of the prairie race caught in the 1940's that penetrated the niceties of innate behavior.

The first experiment, with thirteen of the descendants of Harris' stock, showed no significant preference for the pens in the woods or those in the field. The conclusion seemed evident that after twenty generations of life in a laboratory innate patterns were erased. But Wecker now built two small breeding pens—one in the woodlot, one in the field—where he could raise new litters. The parents were chosen from the twenty-generation descendants of the prairie race who themselves exhibited no preference. When the young raised in the field were old enough to be moved to the long option pen, they unanimously voted for the field. But when the young prairie breed raised in the woodlot were tested, they remained as open-minded as their parents. They did not vote as one for the woods.

What did it mean? One must conclude, as did Wecker, that early experience reinforcing an innate pattern, though unexercised for twenty generations, is sufficient to renew the pattern with all its original force. But early experience that, as in the woodlot litters, runs contrary to the race's evolutionary history has no more than random consequence. Learning, in other words, comes more easily to those with an appropriate evolutionary background than to those for whom an environment carries no evolutionary relevance.

We cannot extrapolate from races of deermice to races of men. But we may recall a variety of comments on animal learning by our foremost evolutionists. Irenäus Eibl-Eibesfeldt, writing of innate mechanisms, has described them as "those reactions and states within the animal on which conditioning and learning depend." S. L. Washburn and David Hamburg: "Evolution, through selection, has built the biological base so that many behaviors are easily, almost inevitably, learned." René Dubos: "The performance of any living organism in a given situation is conditioned of course by environmental forces. But its characteristics are determined by the potentialities and the limitations which the organism has acquired and retained from its evolutionary and experiential past." Sir Julian Huxley, in *Essays of a Humanist*, writing of the behaviorists: "They forget that even the *capacity* to learn, to learn at all, to learn only at a definite stage of development, to learn one kind of thing rather than another, to learn more or less quickly, must have some genetic basis." Or we may recall the sigh of the great neurologist Karl Lashley toward the end of his career: "I sometimes feel in reviewing the evidence on the localization of the memory trace, that the necessary conclusion is that learning is just not possible."

In confronting an environment, the superiority of the individual, of the population, of the race at our stage of human history must rest in large portion on the capacity to learn. But that capacity to learn—which we may grant in *Homo sapiens* as more or less equal throughout human populations—is today founded on biological bases varying through natural selection according to the varying environments in which modern races have evolved. Selective pressures operating on tribesmen in disease-ridden tropical Africa cannot have been the same as those confronting herdsmen on the wind-swept Central Asian steppes. And not only did selection vary according to the natural environments we faced, but also—at least for the last ten thousand years—according to our widely varying cultural patterns which became a part of our environment. All pressures combined to favor biological bases varying slightly or broadly among the world's peoples. And no variance could have been broader or more anciently rooted than in those geographically isolated groups which we refer to as races.

We are permitted, of course, to march to the measure of intellectual fashion and dismiss biological history as having no influence on intelligence. But if we substitute for the subjective term "intelligence" the more measurable and objective phrase "capacity to learn," then we shall come closer to the realities of the human condition. The capacity, the direction, the ease and the objectives of learning must extend and supplement the genetic base evolved to promote survival within a given environment. As biological qualities must vary, so must the qualities of learning.

Throughout all of this chapter we have inquired into *random* inequality of individuals and populations. Inequality has rested on the accident of sexual recombination, on mutation and genetic drift in groups. The inequality of races, if it exists, must be systematic. It must rest on discernible factors in the differing natural selection placed on the hodge-podge of human mosaics to which we give the term "race." Science possesses today no such discernible factors. We possess only evidences of difference.

In the small black race—which I much suspect, from its numbers, to be the youngest of races—we have such evidence of superiority of anatomical endowment and neurological coordination that it must be regarded as a distinct subdivision of *Homo sapiens*. If racial distinction on the playing field is to be accepted, then can there exist theoretical grounds for banishing distinction in the classroom? In the United States the evidence for inferior learning capacity is as inarguable as superior performance on the baseball diamond; yet the question of intelligence remains distinctly unsettled.

In 1966 an enormous Federally financed study, *Equality of Educational Opportunity*, reported on the educational achievements of some 600,000 students in American schools. Referred to usually as the Coleman Report, it became instantly unavailable at bookshops and from normal distributors of scholarly literature. After some eight months of trying, I pried a copy out of the U.S. Government Printing Office. Why it had become unavailable became evident only after long exploration of its 700-plus pages of statistics. The Negro had failed in American schools—failed catastrophically, beyond statistical doubt or sentimental apology, beyond all explanation. It was not a document to be freely circulated in Congressional areas wherein the Negro commanded the swing vote.

I say "the Negro," but let us be wary of typological thinking. A Negro student might exceed a hundred white fellows in academic accomplishment. But up and down and across the statistical tables one found approximately the same population range: curves revealed a repeated 15 percent of black students falling within the same range of academic accomplishment as

77

the upper 50 percent of whites. It mattered a little but not much whether the student had come from a segregated or integrated school. Socio-economic level of the family brought some influence, but small. Worse still for the black was the record of the Oriental-American, subjected in American life to discrimination certainly as rigorous as the Negro, who consistently equaled and in areas excelled the records of white students.

A consequence of the Coleman Report was the hysterically received study by California's Arthur Jensen, published in early 1969 in the *Harvard Educational Review,* suggesting the genetical inferiority of Negro intelligence. It is a persuasive document, so persuasive that there were those who could provide no better answer than to threaten Jensen's life. But the materials must be regarded with care. Are we truly considering intelligence? or a capacity to learn according to the demands of the materialist American environment? The Coleman Report is frank in its description of the tests administered:

> These tests do not measure intelligence, nor attitudes, nor qualities of character. Furthermore they are not, nor are they intended, to be "culture free." Quite the reverse: they are culture bound. What they measure are the skills which are among the most important in our society for getting a good job and moving to a better one, and for full participation in an increasingly technical world.

The Coleman Report offers small comfort for the genuine racist, since its corollary demonstrates that 15 percent of all black students are *superior* in achievement to one half of all white students. But in its record of Oriental-American accomplishment it offers as little comfort for those who regard black failure as a simple product of discrimination. And it offers nothing but unbounded mystification for all those, white or black, who are convinced that equality of opportunity must somehow be presented to all members of a vertebrate society. Whom it must mystify least is the baboon, whose imagination could never encompass the transportation of some millions of human beings from one continental environment in which they had been successful survivors to another offering no such guarantees.

We do not know about race; and that is the final truth today. We know that within a single interbreeding population—let us say the Swedes, or the English, or the white Americans, or, for that matter, one must suspect, the rhesus monkeys—the accident of the night dictates a diversity of intelligence of such order that between 3 and 3.5 percent must be termed feeble-minded. There are between six and seven million mentally deficient Americans. But the occupation of a single geographical area or ecological niche by two populations of differing subspecies, rare in human life, almost unknown in nature, is a man-made situation demanding man-made answers. And until the scientist, without threat to his life, is free to explore in all candor racial differences, and to prove or disprove systematic inequalities of intelligence, an observer of the sciences has little to offer. But then, neither racist nor egalitarian has much to offer, either, beyond emotion.

3. Order and Disorder

The loneliness of man is the loneliness of the animal. We must have each other. The baboon seeks his troop, the bookkeeper his busy office, the buffalo his herd on some far savanna, the weary bricklayer his fellows at the corner pub, the starling his chattering flock among London streets, the hyena his clan, the farmer his wife when the last chores are done, the herring his school in the cold North Sea, all for quite the same reason: because we cannot survive without each other.

The animal cannot stand alone. When Frank Fraser Darling, the Scottish naturalist and conservationist, put forward the brief thought less than twenty years ago, its implications seemed scarcely revolutionary. Yet until that time we had tended to see the societies of animals as accidental, voluntary, disorganized. We spoke of the gregarious instinct. We mentioned the herd, the horde. And, on the contrary, we regarded human societies in a quite different light as a necessity of unique human life with structures and functions peculiar to the needs of men. But as the studies of animal societies progress, we are coming to see all groupings, whether of men or animals, as mechanisms with common mainsprings, common ratchets and bearings, common balance wheels of subtle regulation; for all have evolved to meet the same need, that of one individual for another.

I have offered the definition of a society as a group of unequal beings organized to meet common needs. And there are animal situations where neither the grouping nor the organization may be apparent to the human observer. When Charles Darwin in the midst of his famous voyage made a long stop in Uruguay, he recorded an experience still unsure of explanation. On a huge *estancia* he found a herd of cattle numbering over ten thousand. It seemed to the unpracticed eye a disorganized agglomerate of beasts. But, as all herdsmen knew, the throng was subdivided into groups of fifty or a hundred that stayed always in the vicinity of one another. Then one night came a terrifying series of electrical storms. The cattle panicked, milled, bolted, dispersed in the dark, panicked and dispersed again. By morning it was as if a deck of ten thousand cards had been shuffled and reshuffled the long night through. For the human herders, restoration of original arrangements was impossible. Yet within twenty-four hours not an animal had failed to find its original partners and with them to resume normal social life.

An informed guess today would suggest that Darwin's cattle were directed to find one another by animal xenophobia, the fear and avoidance of strangers. Few forces of order are more universal in animal groupings or more resonant of human ways. It is a basic force of social structure ensuring the integrity of the group, and I can conceive of no other force that so effectively could have sorted out ten thousand cattle all, to the human eye, looking much the same. Yet it is only one of the structural mechanisms making social life possible.

Superficial acceptance of the herd and the gregarious instinct, which one still will find in sociology's literature, failed to inspect the infinite subdivisions lying beneath the mass and yet presenting the true social reality. One recalling Darwin's cattle may turn to more recent naïvetés such as the expectation that black Africa, in our view an enormous undifferentiated herd of dark-skinned people, would willingly unite for rational ends. Now, in our disillusionment, we discover the underlying realities of language, territory, tribe. And one may gain a quick insight concerning values to be gained from the study of animal orders.

Just as the grouping and structure of animals may not be immediately apparent to the observer, so the social function of mediating the problems of unequals may seem likewise obscure. I shall be returning to this at length. But the superb British ornithologist David Lack has left us a record of a case which, while extreme, still illuminates the function. It concerns a blind old white pelican found in the midst of a colony. Pelicans fish in an African lake in a tight, highly organized column, all diving together simultaneously in a comical all-tails-up image. Not for days or probably years had the old blind pelican been able to fish in such a group. How had he survived? They could only have fed him.

The old pelican's inequality was phenotypic, as I described the term earlier. His blindness we must assume was a consequence of experience, whether of argument or of other hazard. Pelican society saved him. But underlying the compulsions of social order one finds not just the disorderly maladventures of living experience, but, far more deeply, the adventures of the night's accident.

The disorder of genotypes—the natural law that sexual conception must produce individuals of diverse potentiality—is the law that dictates the phenotypic order known as society. In other words, among vertebrate species social order starts happening after you are born. As the womb is the cradle of the egg, so is society the cradle of the individual in his life of experience that will reduce, display, or enhance the range of possibilities which accident originally commanded. Society is the evolutionary invention mediating the necessities of order and disorder. But it is essentially a device for the vulnerable.

The solitary life, if indeed it can be said to exist, is a rare sort of blossom on nature's bush. The leopard is such a being. Perfectly armed with fangs and cunning, he needs no assistance on the hunt. Perfectly organized with strength and agility, he needs no partners to protect his kill. When rarely you glimpse him in the African bush, he will usually be sleeping in the fork of a tree with the remains of a carcass as large as himself draped on the limb beside him. No lion or hyena can reach him to make off with his larder; he is as self-sufficient as any creature you will ever see. The relaxation of a spotted foreleg dangling from the fever tree's sulfur-green embrace testifies to leopard freedom; and we may envy him. Yet when the female comes into heat, then he must seek her. And though his social life be limited to a few occasional hours of copulation, still no leopard can afford to live too far away from other leopards.

Few of us can be as lucky as leopards. There are other species, of course, equipped so suitably or subtly to meet life's demands that with minimum social concession they can proceed along chosen ways. There are the badger, the marten, the mink; we may envy a few reptiles like the chameleon or snake. One we need not envy at all is the praying mantis, a delicate if voracious insect with forelegs hypocritically placed as in prayer. It is a species that has made the minimum compromise between sexual reproduction and social antagonism.

The evolutionary dilemma of the praying mantis rests on the eagerness of the female to eat anything, including the male. How one gets by that problem in a sexually reproducing species

I submit to psychiatrists as a knotty one. Evolution, however, has come up with an answer and, grisly though it be, it works. The female is endowed with poorer eyesight than the male, and she can see him only when he moves. But he has the eye of an eagle concerning what she is up to, whether washing her wings or grooming her insect-equivalent of eyebrows before an insect-equivalent mirror. In the time of sexual heat she of course attracts him, but the long history of the species has equipped the male with inhibitions concerning his consort's less attractive qualities. He approaches her, compelled. But a turn of her head in his direction freezes him. He has the capacity to stand without motion, perhaps with one leg lifted, for an hour or more. And, so posed, he remains invisible to her. Sooner or later she will lose interest and return to the diversions of the powder room. And he will make another foray.

That sex can be a dangerous game requires no comment. Back in the 1930's, however, a zoologist named Karl Roeder became so impressed by praying-mantis relations that he made an elaborate study of the neurological arrangements making survival of the species possible. For while the male, with a final leap, may secure himself on her back unnoticed, there to proceed with copulation, a fair chance exists that he will fail. At the last instant she will glimpse his movement, seize him with her deceptively named forelegs, and begin to eat him. She begins always by chewing off his head.

Having lost his head, he literally loses all fear of her. Roeder found that the center of inhibition lies in the male's brain, while the sexual drive has its center in the abdomen. Headless, abandoned freely to the sexual compulsion, he wrenches from her grasp, mounts her back where she cannot see him, and copulates. Then slowly he will weaken, lose his grip, die. When he slides off to the ground, she will discover him again and finish eating him.

I am aware of no species in which the sexual necessity has commanded arrangements quite so unsatisfactory for the male. In general, the same sexual reproduction that brought to natural selection the disorder of genetic diversity brought the need for a degree of order between the sexes, however brief. There is a family of flies called *Empidae* in which the male brings the cannibalistic female a gift, a bit of food wrapped in a web. While she unwraps it, he copulates. There is even one species in which the male, whether out of laziness or rudimentary cynicism, fails to include the food in the package. It still works.

While it might be said that whether we consider leopards or badgers, the praying mantis or flies bearing gifts, we deal with the minimum society that sex commands, still it does not fall within my definition of a group organized to meet common needs. A population of solitary animals must exist in such a reasonable vicinity as to make contact when reproduction is necessary. But this is the population or the deme that genetics describes, an isolated or semi-isolated grouping of individuals with genes sufficiently co-adapted to offer a reasonable probability of viable offspring. Its one organization is genetic, and what we observe is simply the embodiment of the gene pool. It is not a society.

There are other groupings which I shall not include as societies since they lack organization. The more we study herds, the fewer we find lacking truly social, organized subdivisions. Perhaps the vast herds of bison that once inhabited the American West were merely agglomerations; we do not know. Immense herds of zebra, such as one finds in South West Africa, deceive the eye as did Darwin's cattle; they are collections of discrete family parties. Huge schools of fish such as mackerel or herring are, so far as we know, entirely lacking in organization. That they have some need for each other is evident, or they would not school. But what that need is, and what the function of the school may be are questions that have evaded our best students. They are not what I term here societies.

And yet, as the school of fish presents glimmerings of the social necessity, so do such glimmerings appear in beings who existed before sex was born. W. C. Allee was our first great student of animal societies, and at his laboratories in Chicago he found himself impressed by the tendency of sexless, mindless, single-celled protozoa to thrive best in groups of "optimum number." Allee saw it as some forerunner of social life. He discovered that in the presence of a toxic substance, colloidal silver, the protozoa survival rate was far higher in groups than in those individually exposed. Allee could no more explain his protozoa than Darwin could explain his cattle. But he concluded that "potential sociality is as inherent in living organisms as are the potentials of disoperation. This means that social life is not an accident which appears sporadically among a few highly evolved animals. It is a normal and basically widespread phenomenon."

Our need for each other may be as mysterious as the needs of protozoa or mackerel, or as evident as the need of helpless young robins for two parents to feed them. It may be the need of the dangerous, like the wolf pack or lion pride, who cannot hunt efficiently except in a group; or it may be the need of the endangered, such as young penguins who must have adults to defend them from the rapacious skua. Our need may be as complex as that of the baboon who must have his troop not only to defend the individual but to provide an educational setting in which slow-growing young may learn and mature. Whatever the need, the true society, human or animal, encloses structures organizing its members of varying endowment, and fulfills functions which the individual cannot accomplish alone.

We may all of us yearn for anarchic autonomy. And some individuals among even the most social of species may have the superior capacity to go it alone. So we may find a powerful male gorilla high among the volcanoes of the eastern Congo making his solitary way through bamboo thickets. Or we may and probably shall confront a massive, invincible bull elephant tearing apart trees by the Limpopo River, permanently or temporarily enjoying his anarchy and the delights of a bad disposition. But we shall never find an elephant cow or a female gorilla bathed in such luxury. Vulnerability, whether her own or that of her young, will keep her securely in the social embrace.

Out of our need for each other we form our alliances, forswear our temptations, accept our compromises, obey happily or unhappily the rules of social order. The human being, observing the species, will recognize our common plight. The order of the gene pool is sufficient for the genotype with its demand for diversity. But for the development of the phenotype in confrontation with environment, such order is seldom enough. And so evolution has presented us with organized societies as the survival mechanism for the vulnerable. And we shall see, I believe, that man—the stroller in space, the presumable master of all he surveys—is in truth as vulnerable a creature as nature has ever produced.

2

Defense consumes the major portion of most national budgets, just as it consumes the major energy of most animal societies. So fundamental is the obligation of the group to protect the individual that the alarm signal, in almost all species, becomes the very criterion for society itself. Lacking an alarm alerting its members to danger, the animal group, with very few exceptions, may be described as no society at all but a simple aggregation.

The alarm call may act simply as an alert, or it may set in motion the most complex of defensive actions. We are most of us familiar with the dense flights of starlings and the capacity of the flock to maneuver as one. How the birds communicate their directions for sudden turns, swoops, swerves, we do not know and perhaps never shall. But Oxford's Niko Tinbergen, who with the celebrated Konrad Lorenz pioneered in Europe the new science of ethology, has explained through natural selection how the birds have acquired their capacity.

The starling's most dangerous enemy is the peregrine falcon. Unlike the goshawk, which with a dazzling swoop takes its prey off the ground, the falcon strikes in mid-air. So distinct are the modes of attack that the sophisticated crow, looting a newly sown field, will take flight when sighting a goshawk; sighting a falcon, he will crouch. For the starling flock, airborne in countless number, there would seem to be no defense against the falcon. But the alarm call wakens miracles.

So incredible is the speed of the falcon's dive that at the moment of impact it may be moving at 150 miles an hour. To survive such an impact the falcon is equipped with powerful legs and talons, but its vulnerability must be obvious. At such speed, should anything go awry and the falcon strike its prey with a wing instead of its talons, the wing will be broken. And so, through the ages, starling and falcon have perfected their relation to the nicest of balances. Sighting a hovering falcon, the most alert member of the starling flock gives the instant alarm call. Now all crowd together nearly wingtip to wingtip and begin their unpredictable mass maneuvers. There is no leader. Somehow all participate in a common sense as to what the next swerve or dip will be.

The falcon of course dares not dive. So perfect is the starling social defense that any attack would be suicidal. And so he waits, hovering. And natural selection takes its toll. If one of the hundreds of massed birds has weakness of wing or fails in some fashion to sense the next turn, then it is separated from the flock and the falcon has it. And the unequal starling will leave no further descendants to encumber the flock's gene pool.

Starling defense illuminates the value of society to a group of unequals. In any given situation of danger one member, whether superior in alertness or simply lucky, will be the first to sense threat. And with the alarm call, the perception of the first becomes the property of the group. Spotting a golden eagle, a Wyoming prairie dog will give a specific cry and all members of his town will dive into their burrows. Spotting a coyote, he will give another, and all will stand on their mounds, alerted, watching. The alarm call is the social mechanism ensuring that all but the dullest or most careless will be prepared for the common need.

While in many an animal arrangement, such as the starling flock or the prairie-dog town, the giving of an alarm exposes the individual to no special danger, in others it may entail high risk. Animals benefiting from concealment, for example, give away their position as they give the alarm. Solutions may take such turns as camouflage. Some years ago the young Rhodesian scientist C. K. Brain and I were looking for samango monkeys in the high forests of the Vumba Mountains, a range lying between Rhodesia and Mozambique. The samango monkey is arboreal, brown and rather large, with the creased, lamenting face of some ancient Jew. At a distance, a troop is not too difficult to spot, since the heaviness of body makes the forest canopy dip and snap with their leaping about. When the intruder comes beneath the trees, however, all action ceases and the monkeys become difficult to find. There is silence, broken only by the occasional *tweet-tweet* of birds. My more knowledgeable companion pointed out that the *tweet-tweet* was the samango's camouflaged alarm call.

Elephant camouflage is even more subtle. In the earlier years of animal-watching I had frequently been puzzled by how my presence would attract the attention of all members of an elephant party simultaneously. There seemed no signal, audible or visible, nor did individuals, so far as I could see, take their cues from each other's movements. The problem, I found, had bothered others less innocent than I, and Irven Buss had solved it in a European zoo. There the keeper had led him to an elephant cow and directed Buss to put his hand beneath her throat. He could hear nothing, but he could feel a vibration like a kitten's purr. An undisturbed elephant party in a Uganda woodland makes a communal purr which they can hear and we cannot. With any disturbance the purr stops. The elephant's alarm call is silence.

Few species have evolved such subtleties, but many have evolved variations. The avocet has one call for the herring gull, its most dangerous predator, another for all others. Song sparrows have three stages of alarm, white-tailed deer four. Despite the unshakable convictions of reinforcement theory, and B. F. Skinner's personal verdict that "an animal must emit a cry at least once for other reasons before the cry can be selected as a warning," alarm calls seem innate. Baby blackbirds, hungry and begging in their nest, will go silent at the parent's alarm. Tinbergen has demonstrated that a herring-gull chick, still inside its unbroken egg, makes begging squeaks which it will silence when the mother gives her alarm.

While the alarm benefits the individual, in a sense it is the property of the group. That the impulse to give the call must have some innate basis becomes most evident when the individual giving the alarm decreases thereby his own chances of survival. An incident that my wife and I observed in the summer of 1968 well illustrates the action that for many biologists represents a paradox in natural selection.

The impala is an antelope little larger than a gazelle and, besides being one of the most beautiful animals in Africa, is one of the most defenseless. While the male carries his swept-back, lyre-shaped horns with splendid defiance, as weapons they are toys for children. In southern Africa the impala is the lion's favorite dish, just large enough to be worth taking, just small enough to demand a minimum of exertion. And so they gather always in herds where all may benefit by the long snort of warning given by the first to sense danger. But that warning may entail a voluntary acceptance of risk.

One afternoon we passed an all-male herd of twenty or twenty-five browsing impala, then a few hundred yards away came on a lioness sleeping under a tree. Approaching her too closely, we disturbed her. She rose, and for the first time was observed by the impala. We could hear the instant far-off snort. Now the lioness moved away at deliberate pace toward another tree and another spot of shade. Immediately two impala detached themselves from their fellows and came running after her, sending back to the party repeated warning snorts. When she settled herself again, the two still lingered, watching tensely, giving their occasional snorts. Not until she had most evidently gone back to sleep did they trot away to rejoin their fellows, who never for a moment had stopped eating.

One cannot say that the two impala had accepted risks of a suicidal nature by following a lioness as sleepy as this one. Nevertheless, one can state in very nearly mathematical terms the survival value of approaching or fleeing the presence of a lion of unknown antagonism if you are an impala. The mathematics take on a dramatic tinge if you are a Thomson's gazelle, with a warning system quite different from the impala's snort.

The Tommie, as it is known throughout all of East Africa, is the tourist's delight. Their numbers are uncountable—in the Serengeti plains at least half a million. Smaller than the impala, they weigh rarely over fifty pounds. Their dark eyes are enormous, comparable only to the paintings of children by Marie Laurencin and her descendant candy-box Paris painters. The Tommie has the additional tourist advantage, which candy-box children do not, of a short little tail wagged in the most ingratiating of circles. But outweighing all other qualities is its supreme disadvantage of being delectable. Tommie meat has been the favorite of human hunters so long as we have hunted. And for far longer than that the little gazelle has been favored by the entire community of savanna predators, even by the lion when no larger game is about. The wonder is that the Tommie so thrives, and perhaps the behavior called "stotting" has contributed to the species' success.

To stot is to warn your fellows of an approaching predator by leaping stiff legged on all four feet like an animated jumping jack. You can reproduce the behavior by taking thumb and three fingers and bouncing around on a table. So conspicuous are the leaps that a stotting Tommie will spread his warning around a mile of savanna. Yet no alarm could be so dangerous, for if you are being approached by a pack of hunting dogs, its leaders hitting forty-five miles an hour as they come over a rise, then while you spread your alarm you are losing ground at a menacing rate.

There is a temptation to regard a stotting Tommie as stupid, so suicidal may be its situation while spreading the alarm. Were we watching a human situation, we should speak of heroism, but the concept is anthropomorphic as applied to gazelles. What can impel an animal to reduce sharply its own chances of survival, even though the chances of the group be as sharply enhanced?

The question strikes at the heart of older assumptions concerning natural selection, assumptions held by many biologists even today. Since Darwin's time we have seen selection as a process choosing between individuals, the fit from the less fit. A stotting Tommie, in such terms, would be selected out and any such innate tendency in individuals would vanish from the species. Yet we have here on these sweeping East African savannas a superbly successful species in which the trait, far from having been selected against as maladaptive, has become species-specific. The older view of natural selection, formed before population genetics, cannot be applied.

It was a century ago that Alfred Russel Wallace introduced the concept of group selection—that natural selection in final assessment operates between groups. Wallace, however, was speaking of human groups, in which qualities such as altruism,

93

heroism, sacrifice, mutual loyalty could be valid attributes of a tribe of superior fitness. The problem of the animal remained until 1962, when V. C. Wynne-Edwards published his revolutionary *Animal Dispersion in Relation to Social Behavior*. Wynne-Edwards, of Scotland's University of Aberdeen, is a slim, softly speaking man of utmost distinction, as unlikely a bomb-thrower as ever showed up at an anarchists' drinking party. Yet with his massive book he has emerged as the most controversial figure in the field of evolutionary behavior since Konrad Lorenz first placed his concepts of instinct before a world of hostile psychologists. Unlike Lorenz, however, Wynne-Edwards had so far remained a figure of controversy only within biology itself. Central to his essential challenge is such a statement as this:

> In the case of a social group-character, what is passed from parents to offspring is the mechanism, in each individual, to respond correctly in the interests of the community—not in their own individual interests—in every one of a wide range of social situations.

Selection may take place, in other words, at an individual level of everyday competition; but it takes place also at a community level, favoring traits of benefit to the community's survival. We may think of a baboon troop under attack, and the willingness of the leaders to risk their lives in the troop's defense. We may think of a nation invaded by an enemy, however superior, and the willingness of young males to take ultimate risk in the territorial defense.

In such terms, and only in such terms, can the Tommie's alarm signal be explained as the product of evolutionary process. It is again a matter of the gene pool, like seed lying invisible beneath a cultivated field, and of society, that visible crop ripening or withering in the sun. Through the long ages of competition between Tommies and predators, like crops and the weather, an unseen competition has gone on between Tommie herds. Some produced superior seed. Those populations in which the innate compulsion to give warning ran the strongest most obviously possessed the superior capacity to survive. A quality perhaps once the property of but a few populations spread, through group competition, to become today the property of the species.

Society is the gene pool's testing ground. With fine disregard for individual fate, natural selection must look on the competition of groups for phenotypic evidence as to how genetic possibilities are working. And while you and I may shudder, at first thought, concerning gods so harsh, on second thought we may look a bit closer and glimpse immortality in our mirror.

3

The ultimate object of defense in an animal community is the young. A baboon troop seeking food moves like a naval task force across open savanna and through hazardous bush, with mothers and infants near the center where loom the most powerful adult males, like old-time battleships, while scouting destroyers, the juvenile males, leap and chase along the dubious margins of the open, yellow, quivering sea. A bark of the alarm signal will bring the big males, with their concert of fighting teeth, to face whatever threat may exist. The thunderous evolutionary success of the baboon, however, rests not only on the superb defenses of social action, but also on the quickness and adaptability of baboon intelligence.

The elephant, which has already puzzled us enough, offers through another puzzle a clue to a social function which in some species may surpass the importance of defense. Any elephant-watcher in African woodlands must wonder why such tightness of social order is preserved. There are the all-male parties, of course, which young males join on reaching sexual maturity. These are loose, shifting in numbers and membership. But the cows and their young form the parties of a dozen, fifteen, even twenty that one thinks of as the normal elephant herd. Bulls

may be evident, but let the observer not be deceived. They are usually adolescents, however enormous, who have yet to escape their mother's rule. And that rule is hard. When we come to discuss the alpha fish, we shall inspect the ordered, inviolable discipline of the cows in an elephant family party. A United States Marine would find little to envy.

A question must rise: Why all the order? Surprise an elephant herd at a distance uncomfortable for both observer and observed. While you are turning your car around (there is hardly a professional driver in Africa, facing elephants at anything resembling close quarters, who will not turn his car or at least assess a direction offering fast escape) the elephants will be making their own rearrangements as well. The young will vanish behind a battlement of adult bodies. Since eyesight is poor, trunks will rise to sniff you out. If tusks rise also, the best advice is to place the car in gear. Still one must ask, What has the elephant to be afraid of? Man has hunted elephants for their ivory, it is true, for a good five thousand years. But such subtlety of alarm signal, such social order, seems unlikely to have evolved in such a brief span. In the far past, the massive, extinct sabertooth cat may have been a threat to the young. But today no killer less than man will go near them. If defense has been the selective factor in the evolution of elephant society, then these vast mobile fortresses, marching through a woodland dusk like mountains gone to war, must be a far more nervous lot than they choose to reveal to casual observers.

Defense, one must conclude, while virtually perfect is still a secondary function in elephant social life. What must be the compulsion for order lying on the elephant herd is not only protection but education of the young. As in many another species of higher animal, and particularly those with slow-growing offspring, the command to learn may rank higher than the command to defend, if groups are to survive. When the national budget of a human group fails to accommodate the demands of education, then we may safely anticipate group competition taking its inevitable toll.

A logic ignored until recent years reveals primacy of education in many a species as the selective factor it is: were the accumulations of learning not of supreme survival value to the adult, then slowness of growing up would be maladaptive, a drag on the energies of group or species militating against survival. The elephant has a life span comparable to man's. The male will be almost twenty years old before he escapes the discipline of the matriarchal family school. But there, under the leadership of the oldest and wisest of cows, he will have learned answers to elephant emergencies never to be forgotten in his long life. And such emergencies may be lethal, for the elephant's mass is its vulnerability. If drought threatens his daily quota of green stuff and water, then he must move perhaps hundreds of miles over ancient trails—the elephant roads of African lore—to assured supplies. Man-conducted game reserves may conceal from us the elephant's eternal necessity. But the elephant, denied freedom in his protected state, retains his old-time schooling.

The late, great primate student K. R. L. Hall, of Bristol University, was one of the first to emphasize that primate learning takes place in a social context. The young animal learns not because he is taught, not because as in the laboratories of America he is punished if he fails, rewarded if he succeeds, but because inner necessity drives him to observe, to experience, to remember. He learns much from his mother, and it has been said that a young rhesus monkey knows as much as his mother by the age of one. Yet the great pool of knowledge and experience, exceeding that of any individual member, resides in the group itself.

It was Hall, conducting studies of troops of baboons in a Cape of Good Hope reserve, who released a female baboon brought south from the high-altitude environment of the Transvaal. Foods were different and she had no inkling as to what to eat and what not. Local troops drove her away as they would any stranger. She would have starved had Hall not fed her. Then at last a troop began to tolerate her presence on its fringes. And in two days she learned the entire menu of local diet.

So blocked have we been by the tyranny of reinforcement theory with its rewards and its punishments that today we are only beginning to learn about learning. In a crashing experiment at the University of California, Los Angeles, the iconoclastic zoologist J. Lee Kavanau built a maze with 427 meters of linear runways, 1,205 ninety-degree turns, and openings into 445 blind alleys occupying over 50 percent of the space. Into the maze he turned not the domesticated, highly inbred animals that have furnished us with the supposed proof of reinforcement theory, but wild-caught whitefooted mice. Without pressure of deprivation, without threat of punishment or lure of reward, they learned to run the entire maze not only forward but backward in two to three days. "These activities appear to be the expression of inherited tendencies to explore and develop wide-ranging motor activities," wrote Kavanau. "It is unlikely that these remarkable learning performances even begin to approach the capacities of the animals."

The imposing edifice of education not only in America but elsewhere was constructed before our least knowledge of exploratory behavior in animals. And today, as psychology denies the principle of innate aggressiveness in man, it denies at the same time an innate drive to learn and to master problems of the environment. Washburn and Hamburg have written, "Prolonged youth would have no advantage unless the inner drive to activity led to knowledge and skills." They were writing of monkeys. But if the innate drive in monkeys and mice is susceptible to proof, then conclusions must bear some significance to that supreme learner, man.

The crisis in our schools is larger than the educator, the educated, or the perplexed parent can know. Our beleaguered species may face student demonstrations, sit-ins, protest marches, campus violence, school vandalism beyond explanation. Yet we face not only the flaming torches of a new flaming youth but the frozen orthodoxies of educators as ignorant of advances in learning theory as was Neanderthal man of the planted field. And the environmentalist does his considerable best to protect the environment of ignorance.

In the United States the most intelligent, uninhibited laboratory of primate research has been that of Harry Harlow's training ground at the University of Wisconsin. Harlow is an erratic genius of American psychology who, departing at wide angle from behaviorism's mainstream, has led a professional life, I must assume, of quiet desperation. It was he who first demonstrated in the chimpanzee what all ethologists today accept as exploratory behavior. So tall does Harlow stand that to ignore his conclusions becomes impossible. Yet academic fashion has managed to treat demonstrations from Wisconsin with sufficient selectivity so that minimum harm is done.

An early triumph concerned sex. For the environmentalist, denying all innate human compulsions, a thorn bush of a problem is the sexual impulse. But in 1958 from Wisconsin itself came a paper implying that sex is learned. It was by William A. Mason, whose later study of the callicebus monkey I have cited in another book. What Mason presented to the American Psychological Association was an experiment with six rhesus monkeys raised in total isolation and a single wild rhesus caught at the age of twenty months when still sexually immature. Not one of the isolates, deprived of experience, ever achieved sexual effectiveness, whereas the wild monkey matured in normal fashion. The paper became a milestone in environmentalist thought. The doctrine that man is nothing excepting what he learns disposed even of sex. Despite all that has happened since 1958, the paper is still quoted.

None of us—certainly not I and, I am sure, not Mason—understood in those days what Hall was later to describe as learning in a social context. But something seemed odd. In 1960 on a visit to Kampala I discussed the perplexity with Niels Bolwig, then lecturer in zoology at Uganda's Makerere University. Bolwig shrugged. He himself had raised two baboons together from birth, one a male, one a female. They had had no contact with adults, yet when the time came for sex they had copulated cheerfully and competently.

What was wrong? The following year Harry and Margaret Harlow, in *Natural History*, published a long record of their efforts at Wisconsin that seemed to confirm Mason's experiments on a grander scale. The Wisconsin rhesus colony had been established in 1953. No less than fifty-five infants under twelve hours old were placed in isolated cages and treated like human babies. They thrived so well that the mortality rate was lower than among those raised with mothers. The Harlows intended to turn all into the general breeding stock, but the intention came to nothing. Not a monkey on maturity ever reproduced.

Another ninety, raised with surrogate mothers—mechanical substitutes of one sort or another—came off little better, though a few reproduced. The answer seemed to lie with some bond of affection between mother and infant laying the ground for later sexual success. This was good enough for the environmentalists, and the Freudians too, and the paper was widely quoted. One could not but be haunted, however, by Bolwig's motherless baboons. And, though overlooked, there was in the Harlow paper more than a hint concerning the importance of play to the young animal.

In November 1962 the Harlows published the paper that from our social view may be regarded as definitive. The importance of the mother—apparently critical to the emotional development of the infant—had dominated the preceding paper. Now with the ripening of still another experiment, all earlier fallacies were revealed. The mother remained significant but dispensable; society was compulsory.

The Harlows' large group of monkeys raised in isolation—animals such as those which had proved to environmentalism's satisfaction that even sex is learned—were by now five to eight years old, and neurotics to the last monkey. They lived in apathy. For long periods a typical movement would be to clasp hands over their heads and rock back and forth. Compulsive masochism was a behavior norm: some pinched their own skin in the same spot hundreds of times a day, or chewed and tore at their bodies. The Harlows in 1960 tried group therapy and put nineteen on the monkey island of a Wisconsin zoo. Fighting at first was severe, but in time things settled down to a relatively normal social system. With the return to the laboratory, however, all improvements disappeared. After two years more the Harlows concluded that the effects of youthful social isolation were irreversible, and that therapy had no lasting value.

In the meantime, their definitive experiment matured. Four infant rhesus monkeys were reared with their mothers in such fashion that the infants had access to one another and normal play. This was the control group. As might be predicted, the infants from the start had lively, uninhibited social relations, at first exploratory, then rough-and-tumble, then flight and pursuit. While even in the first six months brief sexual posturings took place, they became even by the end of the first year more frequent and adult in form. Throughout the second year, even though the monkeys were still immature, a full repertory of adult sexual behavior developed. These young ones, however, still had access to their mothers.

But the experiment also concerned two groups of four infants each, all motherless. From birth they had no contact with adults, no possible access of adult imitation. The Harlows constructed a playroom which artificially reproduced the normal environment of the rhesus monkey. It included climbing and swinging devices, a ladder, movable toys, even an artificial tree. For twenty-three hours and forty minutes of every day each infant was isolated with a surrogate mother built of wire and terry cloth, the device which had so spectacularly failed to encourage the development of normal young. But for twenty minutes a day the infants were allowed access, in groups of four, to the playroom and to each other. Their start was slow. For two months they spent most of their time clinging together, or moving about in a tight group. Then gradually relations relaxed. The same development as among the mothered infants unrolled. By the age of one year their relations and developing sex-play were no different from those of the privileged control group. "No member of the group shows any sign of damage by mother-deprivation."

Adult example offered no least model for the learning of these monkeys. "It seems possible—even likely—that the infant-mother affectional system is dispensable, whereas the infant-infant system is the *sine qua non* for later adjustments in all spheres of monkey life." The social context, in other words, is, as Ronald Hall indicated, the imperative of primate learning. And, as the Harlows demonstrated, such learning develops innately even in an artificially reproduced environment with an abnormal limitation of twenty minutes a day for social exposure.

Learning, like oxygen, is something imbibed from the atmosphere about one. Coded in the genotype are the instructions that as living beings we shall seek, explore, try, dare, stretch our developing capacities in relation to our environment—and in a social species that environment will consist largely of our fellows. We may benefit from the experience of our elders, partaking of our society's store of wisdom. Conditioning will of course occur as we learn to avoid the painful, to seek the pleasurable. Nor can the positive or negative influence of parents be denied. But we are not the sole product of the parental relationship, as the Freudians would suggest, nor are we the simplistic, identical ciphers that the behaviorists would find convenient. We are beings created unequal who through learning come to make the best or the worst of our endowment.

An organized society, if it is just, provides a context of equal opportunity for that learning. Sir Arthur Keith once wrote that child-raising is the first industry of every species, and if the industry fails, then the species drifts to extinction. Failure may overtake us through injustice, through misplaced ideals, through carelessness, through apathy. But just as the society unwilling to defend itself is lost today, so the society incapable of encouraging its young to full development is lost tomorrow.

Before we turn to the third major function of society, let us glance back at what modern biology has made available to us so far. There are the forces of disorder. There is the diversity of our conception, restrained only by the potentialities of our population's gene pool. There is aggressiveness coded into our genetic beginnings—in less or greater measure, just as all other characters—that enhances our original diversity by driving us to develop our peculiar endowments. But that aggressiveness—sometimes lessening, more often exaggerating the inequality of beings—threatens chaos and annihilation for social species in which individuals are dependent one upon another. And so the social mechanism enters with its balancing forces of order. The most highly endowed or aggressive must sacrifice a portion of his sovereignty to the group if he himself is to survive. The group offers the protection of numbers, but in such fashion that the superior endowments of the few become the survival guarantee for the many. And finally, perhaps in varying importance throughout the species but of invariable importance to man, there is learning. It may consist of participation in the society's pool of past experience or it may include the anticipation of future experience, but in the just society it is available to every member. And it is a force of order, since it brings to the disorder of unequally created beings the vertebrate law of equality of opportunity.

4

As an ecologist, Wynne-Edwards is centrally preoccupied with the relation of the animal to its environment; and so in his *Animal Dispersion* the major concern with society is its effectiveness at adjusting animal numbers to the resources of the habitat. It is a subject so large that I shall defer any general consideration. In the course of his book, however, he presents a startling definition: "A society can be defined as a group of individuals competing for conventional prizes by conventional means." In

The Territorial Imperative I recorded Wynne-Edwards' definition and suggested that it was so far beyond present comprehension that perhaps it had best be left in the sun to ripen for a while. After four years we may bring the fruit in and see if we can digest it.

Let us recall once again my own definition of a society as a group of unequal beings organized to meet common needs. And then, since the best thing about definitions, like hundred-dollar bills, is to have plenty of them, let us consider still another, that of Stuart Altmann. A Canadian, and one of the younger adventurers pressing into the *terra incognita* of evolutionary behavior, Altmann has already become a world authority in the field of animal communication. And he has done what I have not yet attempted: drawn clear distinction between the population and the society. He defines a population in the classic sense of an interbreeding group the boundaries of which appear at that line where interbreeding drops suddenly to a minimum. Applying the same kind of thinking to society, he defines it as a group of beings in which communication carries understanding. And where communication—a mutual understanding of signals —drops off to a minimum, there one encounters the social border.

It follows from Altmann's thinking that the population and the society may coincide. But they may not. What seems a single society may be divided by racial or religious or tribal lines into two or more breeding populations. So long as they speak the same language—far more significantly in a figurative than in a literal sense—so long as they enjoy the same signals, pursue the same goals, accept the same rules and regulations, all will go well with the large society. But let communication fail, let a gap of misunderstanding develop along population boundaries, so that the signals, the symbols, the calls or upraised tails, the words or phrases or sounds of music no longer carry a universal meaning, then the gap may become a chasm into which with slow horror the whole society may slip and vanish.

Altmann's definition and that of Wynne-Edwards, far from

being incompatible, help to explain each other. For when the Scottish ecologist speaks of conventional prizes, he speaks of goals understood and accepted by all. If you are a male robin, it may be a territory, that patch of land which you defend against intrusion and is exclusively your own. If you are a male jackdaw, then territory may be incomprehensible to you, but the highest rank possible within your flock may be the consummation of heart's desire. And if you are a female robin, of course a male with territory will be your goal, and if you are a female jackdaw, then the attainable male of highest status will be the target in your sexual gunsight. All, male or female, respond to prizes of value understood by all. Yet the prizes are conventional in that they are symbols of worth. The robin's territory may include a certain convenience of food supply. The jackdaw's status may the more readily assist him in acquiring a resounding mate. But the competition is neither for food nor for mates. It is for conventional prizes bearing such value as society disposes.

Wynne-Edwards speaks of prizes "pursued by conventional means." Society proscribes certain means of competition detrimental to its own survival. "Thou shalt not kill" has been a social proscription written not just into the Judeo-Christian commitments but into the commandments of virtually every human community on earth. Unwritten but understood is a comparable clause in the social contracts of monkeys and wolves, of gulls and hyenas; it was understood by the armies of Joshua: "Thou shalt not kill—except members of other societies."

In terms of Wynne-Edwards' definition, the competition of beings becomes the salient premise. Competition must and will take place at all levels of life. From the competition of sperm in the female genital tract to the competition of the aging with the grasping forces of oblivion, from the competition of children for parental attention to the competition of peers in the play yard or sports field, from the competitions of women in dress and figure to the competitions of men for power and prestige we deal with a never ending process common to all species. Nor could its universality be otherwise, resting as it does on the natural diversity of beings and the innate force pressing us to select and fulfill whatever varied potentials we may possess. The Utopian may deny competition as a necessary condition of man. But had we been that pioneer animal lacking the quality, we should never have emerged onto the forest margins of our Miocene beginnings.

Accepting the competition of sexually reproducing beings, then we may understand Wynne-Edwards' view of the social function, as well as Altmann's staking of the social border. Through means of mutual understanding, a group regulates the competition among its members so that we strive for symbolic prizes according to rules and regulations presumably fair to all. Beyond the border where communication fails, however, the prizes galvanizing our society may lie beyond another's comprehension. For East Africa's Masai tribe possession of cattle may symbolize high status, while for the neighboring Kikuyu it is the pursuit of political power and large black cars. Each is slightly mad in the view of the other. The prize of one society may lie in numbers of wives, in another in numbers of acres, in another in the numbers of a Swiss bank account. Within a single society times may change: the lure of the court and the warrior's way, together with a scorn for money and commerce, may give way to the ascendancy of industry and accumulated wealth, which with further tidal changes may in turn give way to the prizes of high management, economic or political.

All are conventional prizes meaningful to social partners, tying society together by mutual ambitions and mutual strivings, even mutual envies. But they must be pursued within social covenants agreed on by all. Here then is the third social function—leading to the cliché phrase "law and order"—which accepts competition as inescapable but by means of conventions preserves the degree of order essential for survival of the group.

The function of society as a playing field governed as in any other sport by goals and rules may seem of less compelling reward to the individual than the functions of defense and education. Yet these are the rules that protect him from the depredations of his fellows, as concerted defense protects him from the depredations of others. And there is more to it than that. Perhaps we might resurrect a hypothesis of innate needs that I projected toward the close of *The Territorial Imperative*.

I suggested that in all higher animals, including man, there are basic, inborn needs for three satisfactions: identity, stimulation, and security. I described them in terms of their opposites: anonymity, boredom, and anxiety. They vary in hierarchy, since the need of the female for security must quite apparently be greater than in the male. The endowment and experience of the individual must likewise dictate varying degrees of this need or that. To a surprising degree, however, security ranks lowest among our needs, and the more thoroughly we achieve it, the more willingly do we sacrifice it for stimulation. So long as we live in a milieu of material deprivation, the illusion that security is paramount will enclose us; and many an error of social philosophy has so been written. But let even a minimum of affluence replace deprivation's demands, and security will give way to boredom, a condition to be avoided.

To a degree as surprising as the low rank of security in the hierarchy of need is the high place of identity. To know who you are; to achieve identification in the eyes of your social partners; to sense a fulfillment of the uniqueness that in truth was once yours as a fertilized egg: I submit that it is the ultimate motive. How many people do you know who, given the choice between fame and fortune, will not choose fame? Only when we encounter and accept the ultimate frustration, anonymity, do we turn down the steps of basic needs to the search for mere stimulation, just as we turn up the steps to search for it when, with security satisfied, we encounter boredom.

Let us for the moment accept the triad of innate needs as a hypothesis and no more. It will take on greater credibility, I believe, and as a hypothesis more reliable predictive power, as we proceed through later stages of this investigation. At this point let us content ourselves with inspecting the psychological function of society in the satisfaction of individual need. For I maintain that a society so designed as to present its members with equal opportunity to achieve identity, stimulation, and security will survive the trials of group selection; whereas one that fails in its psychological function will, in the long competition, be selected out.

Protection of the individual from outward enemies; protection, through education, of every member's right, according to vertebrate law, to equal opportunity for developing his potential; and protection of members, one from another, as through competition they seek psychological satisfaction of their innate needs: these are the three functions of order which any society must provide for its members if individuals and society are alike to survive. They are as true of the societies of any higher animals —and in particular of the family of primates—as they are of that highest primate, man. In non-human societies, through stronger instinctive control, they are in a sense self-operating. In human societies, through neglect or design, they are less so.

From most ancient times we have recognized the demands of outward defense and internal order. They are closest to our evolutionary origins. In more recent times of technological advance we have recognized or begun to recognize the educational imperative, and the inarguable truth that the strongest and most durable of societies will be founded on the maximum development of its members. But not even now do we begin to recognize the psychological function—to glimpse even dimly that as security is enhanced, so are boredom and anonymity. The very conquests of production which today promise to reduce material anxiety from human preoccupation have ignored or repressed other innate needs of the individual.

And we do not know what we are doing. There is no contemporary disaster to compare with the bankruptcy of human reason in its confrontation with the human being. The perceptions of the Elizabethan theater, almost five centuries ago, offered insights more profound concerning the nature of man.

In simple bewilderment we watch the spread of violence through what once were peaceful streets. We note in anguish the rise of crime unprecedented in America; and we blame it on our racial problems while ignoring its rise in lands where race is no factor; or we blame it on poverty, forgetting that in the 1930's, when poverty was a common possession, crime was endurable. Earnestly we grope for clues to explain the revolt of the young, the persuasions of alcoholism, hallucinatory drugs, pornography. Explanations become dust even as we touch them. Yet why should such a simple explanation elude our reason?

The hungry psyche has replaced the hungry belly.

5

Jean-Jacques Rousseau at the beginning of *The Confessions* wrote, "I am made unlike anyone I have ever met; I will even venture to say that I am like no one in the whole world. I may be no better, but at least I am different. Whether Nature did well or ill in breaking the mold in which she formed me, is a question which can only be resolved after the reading of my book."

That all men are different; that nature, through sexual recombination, breaks every mold in which men are cast: neither comment detracts from a mad night's accident that could present to mankind a Rousseau. His autobiography, one of the most enthralling books ever written, with breathtaking conviction confirms the assertion. He was a variant of the first order, a genius scorned, humiliated, pursued, enshrined, a loner whose influence can be compared to few since the time of Christ. His influence on our own time can be compared to none.

Rousseau conducted his passionate meditations a century before Darwin, yet he saw man with equivalent clarity as a being
rooted in nature. He distrusted democracy yet bequeathed to
the world the principle that government is the servant of the

governed. No atheist, still he dethroned God and placed human destiny in the hands of society; both Catholic and Protestant churches responded with equal savagery. No man to be cowed, he spoke against tyranny and for the sovereignty of the individual as none other in his century. And yet, with *The Social Contract*, he presented what was scarcely a contract at all but a document of surrender of the individual to the group. When Dobzhansky writes, "The philosophy of modern democracies, of Western and Eastern varieties as well, is the doctrine of equality, natural goodness, and the limitless perfectibility of man," with a certain reservation all is purest Rousseau. Perfectibility was to be the contribution of the nineteenth century. But as the image of God may be viewed from a thousand angles, so the image of Rousseau inhabits our day.

And the image well may inspire the future as well. Should any doubt it, they have only to read a roaring summation by the Durants:

> First of all, of course, he was the mother of the Romantic movement. . . . But what shall we mean by the Romantic movement? The rebellion of feeling against reason, of instinct against intellect, of sentiment against object, of solitude against society, of myth and legend against history, . . . of emotional expression against conventional restraints, of individual freedom against social order, of youth against authority, of democracy against aristocracy, of man against state.

The Durants' catalog of revolt, one image of Rousseau, forms a useful checklist for measuring his reincarnation in the spirit of the contemporary young. With the ebb of the Romantic tide many an idol was carried out to sea, never, we thought, to be seen again. But with the tide's new flow we watch the same old breakers crashing on the same old beaches.

Such a spirit as the Durants record stemmed largely from his earlier work, *Discourse on Inequality*, often known as the *Second Discourse*, which he published in 1755. In it he analyzed primal men as equal and good, the coming of property rights as the source of inequality, and society as the final instrument of ruin created to perpetuate inequality. In 1762 he presented *The Social Contract* as the revolutionary society in which, property abolished, individuals surrender all sovereignty to "the general will," thus regaining as fully as possible the amity and equality of their origins. The principle was to reappear as the *mystique*, if not the reality, of the totalitarian state.

Rousseau's work appeared over a century before Darwin's *Descent of Man*, whereas mine appears just a century afterward. And if I have taken his title and dedicated this work to his memory, it is to throw into sharpest relief just what the natural sciences have brought to our understanding of man and the group. In many a way his mind was remarkably modern. He saw man, as I have mentioned, as a portion of nature. He looked to human origins for better understanding of the human outcome. From many a hint one may gather that he pondered over the way of the animal as of significance to the way of man, and one must bow to a visionary who centuries before the coming of ethology glimpsed a truth. And finally one must recognize that Rousseau's objective, no less nor more than my own, looked to nature and natural law for human solutions.

In a sense he asked all the right questions. But he asked them too soon. Without the theory of evolution to guide him, without the past century's assimilation of proven conclusions in the natural sciences, and in particular without the explosion of the past two or three decades that has transformed biology into virtually a new science, Rousseau could use only his intuitions concerning the nature of nature. And never could a man have guessed so disastrously wrong. In *The Second Discourse* he saw primal man as

wandering in the forests, without industry, without speech,

without domicile, without war and liaisons, with no need of his fellow man, likewise with no desire to harm them, perhaps never even recognizing anyone individually.

Fifty hens in a barnyard flock know each other individually. Even Darwin's cattle would be skeptical.

Males and females united fortuitously, depending on encounter, occasion, and desire. . . . His appetite satisfied, the man no longer needs a given woman, nor the woman a given man. . . . One goes off in one direction, the other another, and there is no likelihood that at the end of nine months they have any memory of having known each other.

Skua pairs, throughout the black Antarctic night, separately circle the continent, yet find and know each other when the winter night is gone.

It is impossible, in that primitive state, that a man would sooner have need of another man than a monkey or wolf of its fellow creature.

Rousseau's vision of asocial primal man became his founding fallacy. He could not know that not a species in our primate family since the early pro-simian, the mouse lemur, has led a solitary life. He could not know that life in organized societies is so characteristically the animal way that a few brief references to such species as the leopard have been sufficient to dispose of those capable of solitary existence. He could not know that xenophobia in a state of nature is as common as the grouping of familiars.

"Man is born free, yet everywhere we see him in chains" is the celebrated opening line of *The Social Contract*. Yet more definitive of Rousseau's thought is the opening line of *Emile*, published the same year and a work he regarded as more important. Since critics are frequently accused of distorting his meaning in translation, I quote it in French:

Que la nature a fait l'homme heureux et bon, mais que la société le déprave et le rend misérable.

With the opening sentence of *Emile* the Age of the Alibi was launched: Nature made me happy and good, and if I am otherwise, it is society's fault. Rousseau's founding fallacy that primal man knew no society is compounded by the second assumption that man in a state of nature was happy and good. That Rousseau knew nothing of the territorial imperative in animal life and regarded the invention of private property as the curse man brought on himself becomes a minor ignorance.

What I believe should be stressed is that Jean-Jacques Rousseau in his time had every right to be wrong. How could he know that natural men were created unequal, or that original goodness is as unlikely as original equality? How could he know that the institution of privately defended property, like the institution of society, was an evolutionary invention far antedating man and his whole primate family? How could he know in the days before Dart that man was descended from predatory primates who killed for a living? Not even Darwin knew that.

The catastrophe is not that Rousseau was wrong but that after two centuries we are wrong; that biological advances since Darwin's time have penetrated our thinking not at all; that fashions of thought today are as firmly grounded in the Rousseau fallacies as if the natural sciences had never existed. In his essay on Rousseau the German philosopher Ernst Cassirer wrote:

> All contemporary social struggles are still moved and driven by this original stimulus. They are rooted in that consciousness of the *responsibility* of society which Rousseau was the first person to possess and which he implanted in posterity.

The Social Contract projected that stimulus. Rousseau reasoned that we cannot return to primal man's state; we must accept society as here to stay. But we can arrange a society abolishing property and its consequent evil, a society to which every man abandons himself freely and totally, thereby achieving equality. The "general will" must govern our conduct. And who shall interpret the general will? "It is the best and most natural rule that the wisest should govern the multitude, when there is assurance that they will govern for its welfare and not for their own." And government of the people, by the people? "If there were a nation of gods, they might be governed by a democracy. So perfect a government will not agree with men."

It is difficult to accept Rousseau as a utopian; his belief in his own social contract seems to me more a matter of making the best of a bad job. A Karl Marx might develop the notion of perfectibility, and that such a society would produce a new kind of man. For Rousseau society itself was anathema, and his contract would at least provide an improvement. But the only Utopia lay in the past.

The reality of Rousseau's philosophy is its pessimism, and this seems to be too little understood. Man is a downhill being. Again, from *Emile*: "*L'homme qui médite est un animal dépravé.*" The statement has sometimes been discounted as extreme. Yet in a letter written five years after *The Social Contract* he said much the same: "I am sure that my heart loves only good. All the evil I ever did in my life was the result of reflection; and the little good I have been able to do was the result of impulse."

I must confess my own failure to apprehend Rousseau's pessimism until I encountered the ungentle hand and the unbowed mind of my own Devil's Advocate, Roger D. Masters, the special editor of my choice for this entire volume. A professor of political science at Dartmouth College, Masters is young, slim, with a black beard as formidable as his intellect. Whereas in another generation we had the pioneer thrusting his way through our western wilderness, today we have a Masters pressing through an academic wilderness to establish the biological foundations of political science. His unseen contributions to this inquiry lie hidden on every page. But as a world authority on

Jean-Jacques Rousseau, his contribution to these paragraphs must receive particular mention, as must his limited liability. We are far from being in perfect agreement on all interpretations. But it has been Masters who has revealed to me the similar inspiration investing Rousseau's work and my own. And it is Masters who has pointed to Jean-Jacques' inevitable pessimism.

The organizing principle of Rousseau's life was his unshakable belief in the original goodness of man, including his own. That it led him into most towering hypocrisies, as recorded in *The Confessions,* is of no shaking importance; such hypocrisies must follow from such an assumption. More significant are the disillusionment, the pessimism, and the paranoia that such a belief in human nature must induce. That Rousseau was persecuted for his views, by church and state alike, is a story of horror. But that he quarreled with his friends, suspected those who helped him most, must be the romantic's bitter fulfillment. Exiled from the continent, he was helped to London by the great Scottish philosopher David Hume, who even arranged that Ramsay should paint his portrait. It hangs in Edinburgh today, but Rousseau, infuriated by what seemed to him calculation in the eyes, sensed insult and conspiracy and quarreled with Hume as he quarreled with all. My wife has redrawn the portrait in a fashion that we believe would have better pleased its subject.

I scarcely need state that my own devotion to human rationality is less than limitless. But never would I deny the power of reason to probe and investigate our human ingredients and through comprehension become a force for good. The vision of man as a fallen angel must to me be the most hopeless of all philosophies. A belief in our original goodness leaves the man that we know as far beyond comprehension as he is beyond help. We become inexcusable monsters; and we must accept Rousseau's verdict that the man who thinks is *un animal dépravé.* For such a being, perhaps, there could be nothing but the social contract as he describes it, a total order. Ours are enforced with an abundance of tanks and machine guns.

What Rousseau introduced to modern man, of course, was the philosophy of the impossible. And what must discourage the observer is that after two centuries of invalidation, from the French Revolution to the rape of Prague, we yet do not recognize it for what it is. There remains time, I believe, however, for debate. Is it truly society that corrupts man? Or man who corrupts society? Is there no alternative to Rousseau's total order other than the adolescent dream of total disorder? I myself believe that evolution itself, playing upon us with laws too large for us to see, provides the answer.

Order and disorder are intimately entwined. Without that degree of order which only society can provide, the vulnerable individual perishes. Yet without that degree of disorder tolerating and promoting to fullest development the diversity of its members, society must wither and vanish in the competitions of group selection. What contemporary evolutionary thought can bring to social philosophy is the demonstrable need for structured disorder within the larger structures of order.

I have described Rousseau's *The Social Contract* as not a contract at all but a document inscribing the total surrender of disorder to order. That it violates natural law may be read in its failure to become a social reality without guns and policemen. What we may observe is that the individual has obligations to the group, as the group has obligations to the individual: it is a contract in equity. We shall proceed to explore those obligations more fully, but first we should recognize a paradox. At the heart of the true contract it is the individual who to exist must have the order of numbers; the group that to exist must have the disorder of diversity. By no accident does the tyrant found his powers on the mob; by no coincidence is freedom protected by the constitution of the group.

A contract in equity, delicately balancing the shifting needs proposed by environment's contingencies, is the fluid bloodstream of a lasting social order. We are less than gods, and Jean-Jacques was correct. We are a portion of nature, and again he was correct. But there is no more severe distinction between my social contract and his than that Rousseau's was an agreement between fallen angels, while mine is one between risen apes. His was a charter for downhill beings whose best times lay inevitably in the past. Mine, granted all cynicism, remains a compact of beings on their way up. His was one to prevent absolute degradation. Mine is one that, accepting the human being, accepts likewise a future that we cannot know. But seeks, in faith dubious but no less real, to encourage it.

4. The Alpha Fish

One winter afternoon in New York a woman showed up at my hotel with an animal story. She was a psychiatrist on the staff of a Long Island hospital, and like many another suburban New Yorker she took special delight in winter as the season when one feeds the birds. A little group of seven chickadees, those most impudent of feathered suburbanites, had become her favorites. They would wait for her outside her back door, and scramble with each other as they fed from her hand. Then on a recent morning something went wrong.

It was cold, the ground snow-covered; bushes and trees glistened with ice. Half-a-dozen huddled chickadees were waiting for her, but they would not approach her hand. On such an ice-embroidered morning they could only be hungry. She wondered, was she wearing the wrong colors? What was there in her presence that had enlisted their sudden disapproval? She put the food in a pan and stepped away. Still they would not approach. Then off in the frozen shrubbery there was a flash and a flutter, and the seventh bird swooped down to the pan. Immediately all joined it in a greedy breakfast. Alpha had arrived.

Psychoanalysis is frequently described as the science of the unconscious mind. Sigmund Freud's discovery of unconscious forces systematically contributing to human decision must be regarded—however we may differ on estimates of Freud—as one of the major achievements of science. In 1908 (it was the year I was born: I mention it only as a point of departure for younger readers delving into the fossil past) Freud and many of his disciples met at Salzburg. From that moment what had been the work of one mind became the movement of many. All embraced and continued to embrace the principle of unconscious forces. But before long the pioneers divided, some to vanish in the tortured trails of mentality's jungle.

We might grant the reality of the unconscious mind; but what were its ingredients? Freud, particularly in those earlier years, chose sexuality as the primordial force driving or distorting our hidden processes. For both Carl Jung and Alfred Adler the preoccupation with sex was too much. Jung specifically rejected Freud's insistence on the lifelong consequences produced by the intense sexual motivation of the very young child in its relation to the father and mother. He saw childhood as dominated by fantasy—fantasies with archetypes expressed by all humanity's myths and religions—with adolescence as the critical period in which the inner world of fantasy must make its peace, however painfully, however successfully, with the outer world of reality. Out of such preoccupations Jung left us the words "introvert" and "extrovert."

Alfred Adler took a path entirely his own, along which he very nearly vanished. He has left us as a legacy, it is true—the term "inferiority complex." But he was neither as profound a thinker as Jung, nor as persuasive an advocate as Freud. Perhaps also his psychology brought shudders more and more severe to those students increasingly enchanted by the equalitarian ideal. Whatever was the cause, Adler became as unfashionable as a hat with ostrich plumes. Yet if we are to gain some perspective on the alpha fish, then we must dig up Adler. His concern was not with sex but with power.

Jung and Freud could to a degree be reconciled. Adler and Freud could not. Freudian acceptance of the unpleasant as a state of increasing tension, of the pleasant as tension's release, did not accord with the Adlerian drive upward. Freud's increasing reliance on the life-force arrayed against the death wish likewise fell far outside Adler's preoccupation with power. And there was always the central theme of sexuality and the family which was so to mesmerize a generation of Freudians. But Adler saw all in a social context. He wrote that there are "no problems in life which cannot be grouped under three main problems—occupational, social, and sexual." That the occupational should be placed first among his categories was to the Freudian an absurdity. In his *Problems of Neurosis* he wrote, "Everyone's goal is one of superiority, but in the case of those who lose their courage and self-confidence, it is diverted from the useful to the useless side of life."

A few years later, in 1932, he stated his central thesis with greatest force: "Whatever name we give it, we shall always find in human beings this great line of activity—the struggle to rise from an inferior position to a superior position, from defeat to victory, from *below* to *above*. It begins in our earliest childhood; it continues to the end of our lives." Adler was writing solely about the human being and the main engine driving our unconscious forces. He took no evolutionary point of view. But I will point to my earlier chapter in which I wrote of innate aggressiveness driving any organism to fulfill genetic potential in its responses to an environment. We deal in the animal with the same force that Adler describes in man.

Studies of animal behavior, and particularly those of our primate family which have so proliferated in the past decade, tend to erode the authority of Freud. Sexuality is not what we thought it in the life of the animal, as it is not what Freud thought it in the lives of men. The Dutch ethologist Adriaan Kortlandt wrote: "There is an increasing awareness that classic

Freudian doctrine is a typical product of *fin de siècle* society. The great problem of our time is no longer sex but aggression and fear." Kortlandt's comment, fifteen years ago, becomes more prophecy than statement when one meditates on more recent, more explosive years.

Although Alfred Adler never achieved the elegance of Freud in a sense either scientific or stylistic (and one is tempted to wonder, occasionally, if his writing appeared originally as a syndicated newspaper column), still I suspect that as investigations proceed more deeply into man's evolutionary nature we shall witness his resurrection. Adler is relevant, whereas Freud is not. Adler presents answers (whether wrong or right) for the questions that burn us. Freud speaks for less dangerous days. Above all, it is not Freud but Adler whose conclusions about man are today being reinforced by evidence gathered from that natural world of which man is a part. We may discuss the alpha fish—bird, cow, elephant, monkey, lizard, rat, wolf, cricket, elk, lion, man—as a phenomenon all but universal in social species; we shall find little in the Oedipus complex to guide us. But Adler's central thesis, which I have quoted, could have been written by Konrad Lorenz in *On Aggression*.

The struggle upward, the striving to rise from an inferior to a superior position, from below to above, has as its end product the alpha. There are some social species—usually those gathering in huge numbers like the starling or herring—in which individual leadership is impossible and the struggle does not take place. Occasionally men behave like a leaderless herd, whether impelled by fear or frenzy, but such is abnormal behavior. We must go far to find a human society lacking presidents or kings, chiefs or elders, generals or captains, oracles or prophets, individuals commanding greater influence than their fellows. Claude Lévi-Strauss has doubted that material benefits could ever explain why some people try to be leaders. There are chiefs

because there are, in any human group, men who unlike most of their companions enjoy prestige for its own sake, feel a strong appeal to responsibility, and to whom the burden of public affairs brings its own reward. These individual differences are certainly emphasized and "played up" by the different cultures, and to unequal degrees. But their clear-cut existence in a society as little competitive as the Nambikuara strongly suggests to my mind that their origin itself is not cultural. They are rather part of those psychological raw materials out of which any given culture is made.

The alpha fish is everywhere. A glorious kudu bull, or a rhesus monkey with tail raised to signal his alphaness, he exists wherever there is organized society. As in human cultures, the status struggle may vary—low in the chimpanzee, high in the baboon —and it may also vary even within the societies of a given species. But man did not invent it, and out of the striving upward comes the hierarchy of rank order, which sometimes our lives depend on. Perhaps we should look at its history.

In 1920 the British amateur ornithologist Eliot Howard presented the natural sciences with the concept of territory in animal affairs. In 1922, just two years later, a Norwegian scientist, T. Schjelderup-Ebbe, published in Germany his study of the social psychology of the chicken yard. It centered on his discovery of the pecking order in a flock of hens. From alpha to omega there is a rank order of dominance within the flock, and each hen has the right to peck those below it in the order, while none has the right to peck back. Thus alpha has the right to peck all, whereas none can peck her. And omega, of course, the last in line, gets pecked by everybody and can peck back at none. In just two years the twin principles of territory and dominance, the concepts at present most absorbing for students of animal behavior, came into being.

Howard, despite his study of innumerable bird species, was conservative in confining his conclusions to bird life, the world he knew. Like Howard, Schjelderup-Ebbe went on to study sparrows, pheasants, ducks and geese, cockatoos, parrots, canaries. He was anything, however, but conservative. "Despotism," he wrote, "is the basic idea of the world, indissolubly bound up with all life and existence." He went beyond life: "There is nothing that does not have a despot. . . . The storm is despot over the water; the lightning over the rock; water over the stone it dissolves." He even recalled a proverb that God is despot over the Devil.

Getting carried away with a good idea is human, forgivable, perhaps even necessary if anyone is to listen. An immediate consequence of Howard's modesty was that no one heard about territory for years. But there is little doubt that the man who discovered the pecking order got carried a bit too far and succeeded only in giving his discovery an unattractive odor. When in later years continental ethology began to take form, Konrad Lorenz alone seems to have been impressed by the significance of dominance in social structure, and some of the most memorable stories in his *King Solomon's Ring* bring us the intimate details of the status struggle in jackdaw life. Yet when as late as 1953 his colleague Niko Tinbergen published *Social Behavior in Animals*, he gave pecking orders less than a page and seems to have regarded them as a peculiar obsession of American science. And it is true that in the 1930's the investigation of dominance crossed the Atlantic and became the particular preserve of three Americans.

It was G. K. Noble of the American Museum of Natural History who gave us the definition of territory as "any defended area." And, while I am not sure, it may also have been from Noble's studies that we have inherited the all-purpose term "the alpha fish." He made many experiments with fish, including the startling one with little tropical swordtails which I reported in *African Genesis* and perhaps should recall here. So obsessed were the males with rank order that though the water in their tank was cooled to a degree where they lost all interest in females, still they continued among themselves an unremitting struggle for status. It was one of the earliest experiments, to be confirmed later in hundreds of species, demonstrating that it is not love, necessarily, that makes the world go around.

In the meantime—and beginning at an even earlier date—C. R. Carpenter's classic studies of wild primates were revealing not only the role of territory in primate social life but what was to become far more important in the studies of many observers, the role of status. Carpenter developed means of measuring degrees of dominance in the gibbon in Thailand, in the howling-monkey troops that he watched for so long in a Panama rain forest, and in the colony of Indian rhesus monkeys that he established on tiny Santiago Island, off Puerto Rico. The measurements revealed nothing so crude as the no-peck-back arrangements of the chicken yard. Rather, Carpenter found rank to consist of varying tendencies of one animal to precede another in privilege and leadership. Although occasionally beta would beat alpha to the fruit tree, far more often it would be the reverse, and alpha's rank would be confirmed.

Carpenter's principal contribution concerned the variety of expressions of dominance. In the rhesus it was severe and relatively clean-cut; he referred to it as a "high dominance gradient." The arboreal howler, in contrast, has a low gradient. The howler has rank, though we might say he seldom pulls it. Out of all these early studies two generalizations appeared which the multitude of recent studies confirm: In any society of monkeys or apes, a degree of dominance among males will be evident, and there will always be one or more individuals with greater influence than others. And in such societies adults will always be dominant over juveniles, males always dominant over females. The gibbon is the only species we know in which the female even nears the dominance of her mate.

Noble extended studies of dominance to fish and reptiles, Carpenter to primates. It was W. C. Allee, however, together with various associates at Chicago, who in the 1930's was making the most exhaustive investigations of Schjelderup-Ebbe's original discovery. The rigid pecking order among chickens was confirmed with only minor deviations. (Years later a study showed this to be no crude peculiarity of domestication. Jungle fowl have precisely the same social order as their domesticated descendants.) But only in the white-throated sparrow could Allee find a pecking order of equal severity. In pigeons, for example, much the same rank order would be accomplished, but, as in Carpenter's monkeys, through probability of dominance. The subordinate pigeon might peck back, but with low chances of winning. And so Allee divided rank order into two categories: the peck-right, in which the subordinate cannot and will not peck back, and peck-dominance, in which he may peck back but will probably lose. And out of it all came the broader concept of hierarchy.

Any society has as its basic structure a hierarchy of members unequally disposed. Perhaps Allee's most fascinating report in his *Social Life of Animals* concerns the studies of the Dionne quintuplets by a group of investigators from the University of Toronto. The quintuplets were identical—developed, in other words, from a single fertilized egg. Yet from an early date they developed a social hierarchy among themselves that varied little as time went by. Such criteria as who pushed whom and who pushed back, who initiated actions and who followed, all went into the grading of hierarchy. And while it is true that the alpha Dionne ranked always highest in mental tests, still the largest and strongest ranked never higher than third in the social hierarchy. Allee wrote: "Whatever the reason, we have come to an interesting, and I think important conclusion, which is that animals with exactly the same heredity may still develop, even at an early age, graded social differences showing that one is not exactly equal to the other."

What, then, is alphaness?

2

The mystery of human personality can no more than a summer storm be confined to logic's prison. Neither the apprehensions of genes in their chromosomes nor the tidy comprehensions of environment's command, neither confessions on a couch nor the conjunction of planets will ever quite explain why you are you and I am I. A leaf flutters quietly from October's bush; a child turns over in bed; footsteps pass beneath a window; the last eye of spark drowses on the hearth. Of such ingredients, perhaps, are we compounded. As we cannot know them, we cannot explain them; as we cannot foresee them, we cannot arrange them. The utopian may in all petulance fling his book on life. The eugenicist, in all gloom, may inspect his navel. The mathematician's cup runs dry. Either life is so much larger than we are, or we are so much larger than life: even this we cannot know for sure. There are mysteries which to some descendant species may be as clear as words on a wall. But you and I, human messengers, deliver poems that we shall never quite understand. We may read them, ponder them, debate and recognize their mysteries, tell them like tales by a fire. Yet something will always lie beyond us. And perhaps it is just as well.

Such is the mystery of the alpha fish. Where does he come from? What makes him tick? We may recognize him, follow him, applaud him, put him to the most useful of purposes. But while many have attempted to describe him, few have been the observers brash enough to attempt to explain him.

Konrad Lorenz, considering his small, crow-like jackdaws, gave as good a description as any of dominant birds: "Not only physical strength, but also personal courage, energy, and even self-assurance of every individual bird are decisive in the maintenance of the pecking order." There is something which at first glance may strike the reader as anthropomorphic in the description; yet I am certain that almost every student of animals will, in general, agree with it. A quality for which we have no other word but "character" is an impression which must be accepted in any description of the alpha. It is something that impressed Fraser Darling in his long observations of Scottish red deer. Two stags will fight only if they are equals. For a long time they will size each other up, then fight or move off. He considered the final estimate to be psychological rather than physical.

We have inherited our brute impression of the master animal from the good old nineteenth-century days of Tennyson's "nature red in tooth and claw." Yet, while might is helpful, it presents no final answer in animal dispositions. C. K. Brain, putting together a captive troop of vervet monkeys in Rhodesia, found that his alpha was smaller and two years younger than the omega. The most obvious characters of the alpha were confidence and a steady gaze. When Brain introduced an older, stronger adult male to the troop, the animal was so badly beaten that within a few days, demoralized, he had to be removed. He died of injuries.

The situation of the vervets well illustrates the rule that size and strength cannot alone guarantee dominance. After a period of observing the famous monkey colonies in Japan in the summer of 1966, C. R. Carpenter described to me a large colony in which one of the three highest-ranking males was thirty-five years old and had not been in a fight in ten years. Kortlandt, reporting on a troop of forest chimpanzees in the eastern Congo, observed a similar situation. The undoubted alpha male, deferred to by the strongest, was at least forty years old.

Seniority may be a quality entering into the alpha, despite infirmities of age, simply because the group has become habituated to his leadership. But a remarkable study of the human being published a few years ago by William D. Altus, while taking nothing away from the virtues of seniority, added much to its mystery.

Ever since the days of Sir Francis Galton evidence has accumulated that eminent men tend to be firstborn sons beyond any factor of chance. No one paid much attention; too many explanations, such as the laws of primogeniture, were available. Later studies of Rhodes Scholars, of Who's Who, of distinguished scientists in America, all showed much the same disproportion of oldest sons. But although primogeniture was a factor of little significance in the United States, still the environmental advantages of being the firstborn offered ample explanation for success. But then came a disquieting moment, Lewis Terman's publication of his analysis of 1,000 "gifted" children. By "gifted," Terman referred to children with an IQ of 140 or higher, representative of the top one percent of the population. And again the preponderance of oldest sons accorded almost perfectly with earlier studies of the eminent.

We know that the IQ test, while designed to measure innate intelligence, is affected to a degree by such environmental influences as the home and cultural pressures. Even so, Terman's findings were incisive enough to be disturbing. But psychologists like Altus had to wait until 1964 to gain a larger sample and more objective criteria to explore mysteries that will remain for a long while unexplained. By then the American experiment known as the National Merit Scholarship had taken such hold as to enlist the young ambitions of all who had ambition at all. Examinations were taken across the land, and from those high-school graduates whom Alfred Adler might have described as supreme in their upward strivings, 1,618 became finalists in the competition. We do not deal with arguable IQ. We do not deal with adult eminence, supported or unsupported by wealth, name, social position. We deal, so far as it is humanly possible, with a democratic competition, open to all, of representatives of a human population of 200 million. And of 568 from two-child families, 66 percent were firstborn. Of 414 from three-child families, 52 percent were firstborn. Even from students of four-child families, an impossible 59 percent were firstborn.

Now, one must inspect such figures with that skepticism which all statistics should invite. We must recognize, for example, that the firstborn may well be the first to enter the competition for National Merit scholarships. The younger ones may not yet have had the chance to prove themselves. And one must recall the environmentalist argument that the oldest child has advantages of years of parental monopoly, of parental ambitions, of undivided parental pride. But then one encounters a hollow voice from behind the statistical draperies.

The distribution applies only to the 1,618 finalists. To the enormous number of superior high-school students who took the examinations, and who should be comparably affected by such advantages and influences, the distribution applies not at all. Is it, then, intelligence that is affected by seniority of birth? Or is it, beyond that, some quality of alphaness, the upward surge, the drive to achieve? Altus concluded: "Ordinal position at birth has been shown to be related to significant social parameters, though the reasons behind the relations are as yet unknown or at best dimly apprehended."

Seniority, like intelligence itself, has something to do with alphaness, though just what we do not know. Adler called the oldest son "power-hungry," and it well may be. Ellen Cullen, one of Tinbergen's most brilliant Oxford students, made a study of kittiwakes, a distinctive species of gull that nests on cliffs along Britain's west coast. The female kittiwake lays, as a rule, two eggs on successive days, and they hatch in the same order. The sibling older by a single day pecks the younger and becomes dominant. But it need not be so. In herds of thirty or forty African buffalo, seniority plays a part in the ranking. Alpha, however, may or may not be the oldest.

Virility of course was regarded as a primary attribute of dominance when we looked to sex for answers, and an early experiment with ringdoves seemed to confirm it. Two flocks, one male, one female, each had its rank order. In the female flock the omega was injected with the male hormone testosterone, and in eight days rose to alpha. In the male flock the omega, similarly injected, rose as high as second rank. Maleness seemed an undoubted determinant.

Something comparable but ambiguous is suggested by a curious observation by Ralph Masure, working with Allee. In a flock of female pigeons a bird called RY, for her red and yellow marking bands, was omega. For some weeks RY occupied this lowly spot, and then a change got into her, and it was not an injection. What happened to RY they could not guess, but she rose in six days to third rank. Still she was dominated by the top two in the order, but then the birds were allowed to mate. RY did not, and seemed quite unresponsive to male advances. When the time of mating was over and the sexes were again segregated, RY emerged as the solid, unchallengeable alpha among the females. Had she been a bisexual? Later on she was autopsied, but no physical abnormality could be found.

From the normal dominance of male over female throughout species, and from such observations as these, one would be tempted to conclude that masculinity, or even a touch of it, makes for dominance. And if ever there was a species to enhance the temptation and enforce the conclusion, it is that weird and wonderful, grotesque and unfathomable creature—I do not refer to the human being—the over-maligned hyena. From the time of Herodotus his name has been a synonym for cowardice. We have at times even called him a hermaphrodite. Only in the past few years have the legends been demolished. Another of Tinbergen's superb students, a young, large, blond Dutch ethologist named Hans Kruuk, has devoted his days and his nights, year after year, to hyena-watching. And he has found, among other discoveries, that, far from being a coward, the hyena is among the world's most formidable predators. It is the animal's sex life, however, that must concern us here. For while the hermaphrodite canard has long ago been dismissed, still there are elements in hyena sexual arrangements which leave the observer's eyebrows permanently lifted.

In the summer af 1968 I had the privilege of Kruuk's company in Tanzania's Serengeti. When you visit a hyena man, you become, naturally, a hyena guest. And so you find yourself in a Land Rover parked beside a hyena den—a series of holes, probably dug originally by wart hogs, which the hyena clan has appropriated as a nursery—while the young ones bounce and chase each other and adults sleep in the sun. Animal-watching consists mostly of waiting for something to happen. And if your companion is someone who knows more about what you are watching than any other man on earth, then animal-watching becomes the world's most civilized experience. There is the conversation, unpressed, uninterrupted by phone calls, as free as a dream. There is the day in the country with the long swelling African plain below you, the few scattered flat-topped trees like dozing sentinels, the wooded ridge a few miles away enclosing its secrets, perhaps an African cloud or two, long, slim, purple-keeled, patrolling the skies; there is even, like some ultimate gift wrapped in heaven, not the faintest possibility of exercise. And of course there are the animals, playing, yawning, greeting one another when a wanderer arrives. A good day of animal-watching should be poured into bottles, labeled with its vintage, and placed in the cellar, and if you are watching hyenas, then you will encounter a very strange flavor, for as slowly you encompass the hyena facts of life, the more surely will you discover that, like some aging *voyeur*, you are ignoring all before you except hyena genitals.

One of the facts of hyena life is that the female is not only larger than the male, but dominant over him. So rare is such a relation in the vertebrate world that one must suspect it came about through evolutionary quirk, an ancient mutation that, while making no sense, did no harm to prospects of survival. Such an accident befell the phalarope; she wears the bright plumage, enjoys the excitements of defending territory, while he, drab fellow, sits on the eggs. The social consequences of female dominance in the hyena are inconspicuous; but the anatomical consequences are lurid. Even Kruuk has difficulty, sometimes, distinguishing male from female at a distance of forty or fifty feet. Her attenuated clitoris, hanging down inches, can scarcely be told from a penis. And where testicles should be she has an enlarged ball of fat in perfect imitation of the scrotum.

All such observations, whether of hyenas or ringdoves, would point to a profound link, whatever its nature, between alphaness and the male hormone. But we immediately encounter contradictions. The rat, the jackdaw, and many another create rank order while still sexually immature. Green sunfish pursue their competitions when so young that sex can be determined only by dissection. Even accepting that the male hormone may at this age be active, we have those uncounted species in which females and young form their own social groups, excluding males, and live in a female-dominated world. Red deer in Scotland, mountain sheep in the American Rockies, elephants in Uganda, kudu in South West Africa's empty reaches, all have their female ranks, their distinct female alphas, without suggestion of masculine disposition. Neither is there evidence that an extraordinary sexual drive in the male promotes alphaness; rather it seems the other way around. The beta monkey may display no consuming interest in females. Let alpha die, however, and beta be promoted, and his sexual fires will be lighted so that he becomes a copulating hero.

So many variables enter the determination of alphaness that one faces an equation beyond solution. Strength, intelligence, maleness, courage, health, indefinable persistence, ambition, confidence—all are involved. Allee has contributed luck, a neglected factor probably of high significance. But the most remarkable quality—one that would occur to none but the shrewdest observer—is political acumen.

Ethology's accumulated materials are still so scattered that one cannot yet judge how widespread political capacity is as a contributor to the making of animal presidents. It has been observed only in primates, and perhaps in none but our own devious family is wit of such selective advantage that a particular discrimination has evolved. Among primates the political aptitude of baboons received the attention of Eugène Marais, the South African naturalist who lived with a wild troop in the early years of the century. Not until 1969, however, with the publication of his lost manuscript, *The Soul of the Ape*, did Marais' work receive any degree of world recognition. Instead, it was the definitive study of baboon social life by Washburn and De Vore, in 1961, that not only initiated the massive observation of primates spreading like an epidemic through the 1960's, but first called significant attention to this primate aptitude.

Aside from one species, the hamadryas, the structure of the baboon troop has a central group of alpha males who not only defend the whole troop but keep order within it. Washburn and De Vore concluded that the capacity to get along with one's fellows must be a necessity in the alpha male. We deal here not with as mild a creature as the samango or colobus monkeys, but with the most belligerent citizen in the non-human primate world. That alphaness must include the capacity to suppress rugged individualism bears notice. Yet the three or four males of the central group prefer one another's company, act always in concert, never quarrel, and accept any challenge to one as a challenge to all. Were a male however strong, however intelligent, however ambitious, unable to accept these essentially political conditions of existence, he could not be an alpha baboon.

The following year, in 1962, Stuart Altmann became the first observer to draw attention to the capacity for coalition in the rhesus monkey. Among rhesus males, however, such coalitions are not between equals. The steps of rank are distinct, and alpha is an individual who must somehow maintain his rank quite on his own, though he leads a group of animals only slightly less aggressive than baboons. In this situation true political shenanigans have value. Alpha, pressed by beta, will make a temporary alliance with gamma. Or gamma, with designs on beta, will ingratiate himself with alpha. Since greatest conflict comes between adjoining ranks, such temporary alliances may take place at any social level. When Thomas Struhsaker recently published his elegant series of studies of verbet-monkey troops in Kenya's Amboseli, he found that 20 percent of all contacts between males involved temporary coalitions.

From broad enough inspection of animal examples, we can gain some appreciation for the qualities of the alpha fish, but no organizing principles. We may say that dominance occurs. And rank in the social order becomes one of Wynne-Edwards' conventional prizes pursued by conventional means. We may look at armies and governments, political or religious movements, corporation personnel or academic faculties, and we shall find the same invariable structure of rank, the same motivation from below to above, the same command by alphas as in primate troops or elephant herds. In the organization of any society of unequal beings to act as one, evolution has favored the mechanism of hierarchy. But despite his universal appearance in social orders, despite the varied testimony we have gathered concerning the nature of the beast, we still do not know what an alpha fish is.

We have been inspecting hierarchy from the top down. We

have been viewing the alpha as an individual phenomenon. But the dominating cannot exist without the dominated, the leader without followers. Perhaps if we reverse our standpoint we shall penetrate the mystery more deeply. If we take a fish-eye view of the hull of the passing ship, then who knows? Perhaps we shall glimpse the captain.

3

For many Americans then of voting age, life has offered few experiences so exalting as the one that overcame us on a night in 1952 when, grouped around our television sets, we listened to Adlai Stevenson accept his nomination for the presidency. Tears flowed without shame. Voices choked. Hands in the dark gripped other hands, whatever hand was closest. It was as if, half a continent away, a single man struck a hidden bell, and throughout the forests of the American soul, from this tree or that, from the heart of this thicket, from the crown of that lonely pine, bells reverberated in response, bells that had never found voice before, bells unguessed joined in a wild, unpredicted clamor of purpose and hope, of resolve and thanksgiving. The Stevenson movement was born.

I was still living in Hollywood at that time. I had published a novel earlier in the year, and out of consequent bankruptcy had a few weeks before the nomination accepted a film assignment to pay for my folly. The contract involved a quite splendid mountain of hard cash. A few days after the nomination I went to the studio, pleaded a heart attack, varicose veins, and several unexpectedly slipped discs, and we tore up the contract. I joined the Stevenson movement and with many another mad American went to work on the campaign. I became a Follower.

Does the sense of being a follower carry some special reward? One thinks of children, the world over, playing games of follow-the-leader. It cannot be Machiavellian designs of vested interests manipulating the conditioned reflexes of the young so that as adults they will keep their proper places. One thinks of the passage of elephants in silent file; it is a predisposition that makes them so easily trained for the circus. One thinks of elk in Wyoming moving at twilight in long silhouette. One recalls the documentation of single-file antics of sea creatures: marine gars in Florida waters, first the leader, then the followers, leaping over a twig in a glistening stream; enormous mantas, in the same waters, churning in a circle like vast, insane bats; silversides, in Long Island Sound, leaping in a shining, continuous parade over a fisherman's line. William Beebe, off Acapulco, once thought he had found some unknown giant eel. But the eel was a string of fish, moving nose to tail.

A "following response," as ethologists usually refer to it, demonstrates in single-file behavior simply a special case. The newly hatched duckling has a response to his mother's quack, and, following her, learns who his mother is. It is the innate basis of Lorenz' famous imprinting, a learning process that occurs in a few early, critical hours of the young being's life. And it can go wrong in famous fashion. Portielje, for many years the head of the Amsterdam Zoo, hand-raised a male South American bittern who through following Portielje became hopelessly confused as to just who either of them was. When the bittern matured, he took the gloomiest possible view of female bitterns, probably feeling that they were quite the wrong sort. Only after long isolation with a female did he mate with her, and they raised several successful broods. But even then Portielje had his problems, for if he came within sight the male bittern would drive his mate off the nest and perform the ceremony of nest relief inviting the zookeeper to incubate the eggs.

The response may not always be innate. The long-maned hamadryas baboon of Ethiopia is the same animal that Egyptians held sacred and that decorates so magnificently Tutankhamen's tomb. So drastic in the life of the female is the necessity to follow that one would take it as a kindness of natural selection to have made the response instinctive. But it is learned, and learned the hard way in accordance with reinforcement theory. The big male hamadryas has a harem of four or five females who must follow him at a distance no greater than three meters. Should one stray farther, then he leaps at her and reinforces her following response by biting her neck. That following is not innate, but perhaps learned in adolescence, was demonstrated when a female raised in captivity was turned free in Ethiopia. A male promptly annexed her. But no matter how painfully he might bite her, she could not follow him. She did not know how.

We need not turn to those most obvious instances in human life where a following is achieved through the approximate techniques of the hamadryas baboon. We are concerned here, rather, with the subordinate role that is accepted, even sought, with varying responses ranging from the grudging through the docile to the enthusiastic and even ecstatic. For there is a possibility that the response of the subordinate in some fashion illuminates if it does not define the alpha.

Wynne-Edwards, with a penetrating stroke of logic, once wrote that a "hereditary compulsion to comply" must be the real key to social organization. Could none be satisfied with less than alpha, then society would become impossible. But his logic proceeded: any such innate compulsion cannot come about by direct inheritance. Since in most animal groups the alphas, through the privilege of rank, have not only greater access to females but generate in the female more insistent sexual desire, then the alphas, throughout generations, must leave descendants out of proportion to their number. If inheritance proceeded directly, then the capacity for subordination would soon be bred out of a species. But observation reveals no such tendency. Whether we consider mice or we consider men, as Lévi-Strauss pondered his Nambikuara, the proportion of alpha-tending and subordinate-tending remains about the same in every generation. The distribution of the trait, Wynne-Edwards concluded, must be controlled by group selection, occurring by random incidence in any gene pool. "The tendency to comply is renewed in every subsequent generation."

Within us, then, is the upward-pressing force that seeks competition, strives for superiority, and in the normal vertebrate society of equal opportunity allows every individual born his chance to demonstrate alphaness. But were that the only force, then organized society would be impossible and we should have only anarchy. And so there is a seemingly opposite force, a willingness to comply, to accept subordination, and—in a majority of individuals—to accept one's station in life as satisfactory reward for past strivings. But is not the following response a portion of this acceptance? Is there not a successful identification with the alpha that leads to the satisfaction of compliance?

In recent ethological thought, increasing attention has been given to the attractiveness of the alpha fish. The gang of thugs ruling a baboon troop may be feared; but their attraction is such that no unruly juvenile need ever be herded back into the fold. The magnetism of the dominants keeps the troop together. Perhaps George Schaller's monumental study of the mountain gorilla offers as clear a portrait as we shall find of the social attraction generated by a dominant animal.

Schaller's observations were made in the woodlands and bamboo thickets found at an altitude of about ten thousand feet in the eastern Congo's Virunga chain of volcanoes. In the area he observed over 190 animals, almost all of whom belonged to one of nine separate groups. Each group in its search for food ranged over an area of from four to eight square miles. Occasionally a male would wander off from his group for a few days, or a previously unobserved mother and infant would join a group, but through the period of Schaller's study, groups kept a fairly consistent identity.

Group identity had its focus in a total alpha, a silverbacked male. The mountain gorilla is black, but with final physical maturity at the age of ten the silver appears on his back. In Schaller's largest group of twenty-seven individuals there were four silverbacks, but only one held command. When he rested, the group rested around him, with adolescent blackbacks frequently on the periphery. All seemed constantly aware of his movements, or took their cues from those close to him. When he moved, all moved. If he moved slowly, the group too might only gradually get under way. But if he signaled a change of activity with a rapid, stiff-legged gait or a characteristic succession of grunts, all might well run to gather in his wake.

Ethologists generally agree that it will take new and larger concepts of dominance to explain both the fear and the attraction generated by the alpha. In late 1967 M. R. A. Chance, one of Britain's most thoughtful primate students, contributed such a concept with a paper on "attention structures." Chance is at Birmingham, and has devoted much of his career to reflection, an activity sorely needed at this stage of ethology's rapid advance. His own attention was attracted by Hans Kummer's studies of the hamadryas and the intense concentration lavished by the female on her overlord's actions. When we recall that she will get her neck bitten as a reward for distraction, the concentration seems explicable. But almost no punishment or even threat enters into the relation of the silverback to his gorilla followers. And attention is as perfect.

"Broadly speaking," wrote Chance, "dominance is at present considered to be that attribute of an animal's behavior which enables it to attain an object when in competition with others. . . . A more rewarding way of defining the dominance status of a supremely dominant individual is that he or she is the *focus of attention* of those holding subordinate status within the same group." What is important among gorillas, in other words, is not so much that the leader leads but that the followers follow.

We are now looking at the alpha fish from a properly fish-eyed viewpoint. And perhaps we may begin to understand why it is so difficult to describe alphaness in terms simply of age, of intelligence, of physical strength, or even of such qualities as persistence or determination. While we still merely penetrate more deeply the mystery of personality, may it not be that true alphaness rests on the capacity to attract and satisfy a following?

When in 1969 the Canadian social anthropologist Lionel Tiger published his astonishing book *Men in Groups*, he introduced to sociology the principle of the male bond, the propensity of males throughout all of *Homo sapiens* to gather in groups that exclude the female. In the course of his discussion he turns to the relative failure of women to achieve high political office unless their lives have been associated with men who achieved it at an earlier date. Despite all the advances of women's suffrage and the power of women's votes, little has happened. Immediately after female suffrage was introduced to Japan, for example, thirty-five women achieved national office; there are now but eleven. In Canada and the United States, women are far less likely to vote for women than they are for men. In no parliament are there over 5 percent women. The Supreme Soviet may contain 17 percent women, but it likewise retains no power; the Politburo contains none.

Tiger makes a suggestion radical in sociology: "That females only rarely dominate authority structures may reflect the female's underlying inability—at the ethological level of 'pattern-releasing behavior'—to affect the behavior of subordinates." To meet the outrage of feminists there is perhaps no answer but hiding in a cellar. For a resident of Rome to assure female militants that women walking down the Via Veneto, the Spanish Steps, or even through St. Peter's itself suffer not the least handicap concerning pattern-releasing behavior in the Italian male is an argument which I suspect will forward one's chances of survival not a whit. Yet one must ponder.

Admittedly, the following response of the Roman male is not precisely of the order that Tiger discusses. Yet the historic capacity of the female to command attention within a sexual context makes all the more notable her incapacity to command it within an authority context. Why should it be? Is the male bond so strong as to dictate failure? In few primate species is there a bond so strong, yet no monkey or ape society accepts female leadership. Is it possible that biology has arranged that the vulnerable primate female with her slow-growing young—and certainly no less our own ancestral hominids—herself normally possesses following response only for the male?

Whatever the answer to male and female, the organizing principle of true alphaness seems to be Chance's attention structure and the satisfaction of following. And we are presented with a criterion for social failure. We may think, for example, of all those vertebrate societies in which, with the few exceptions I noted in an earlier chapter, an opportunity to rise to alpha is available to all males. Now, the stability of animal social orders, in contrast to human, is a characteristic almost as universal. Jockeying for rank goes on, along with occasional exchanges, but almost nothing resembling human revolution takes place. In the past we should have dismissed this as a simple difference between animals and men. But is the dismissal valid? In a society of truly equal opportunity, likelihood dictates that the best-qualified alpha afforded by the group will be the animal to assume the role. He will indeed be the animal of most powerful charisma, the one with the broadest capacity to induce a following response in the entire group. Stability is the consequence.

But now let us think of human societies, with their built-in guarantees of inequality of opportunity. The true alpha may rise; or he may remain in the shadows of anonymity. And in his place may appear the pseudo-alpha lacking any reliable capacity for inducing a following response. He has gained his authority, he must maintain his authority, and to secure social stability he must exert his authority through intrigue and power alone. And no road to ruin lies so open as when authority and alphaness fail to coincide.

It is a thesis which I shall return to in my next chapter, when we turn to the omega, and in particular to the pain of being young. But before we become deafened by the yells from below, let us assure ourselves that the alpha fish has the value to man which evolution proposes.

4

Contemporary studies of primates in the wild began in Japan in the early 1950's, though western science knew little about it. Observations were published in Japanese. The few translated papers to penetrate the language barrier attracted small attention. Japanese science was so far ahead of the scientific world-clock that monkeys and apes had not yet become fashionable. But when ten years later primate behavior, like some animated mini-skirt, suddenly arrested the attention of western science and public alike, the Japanese Monkey Center found itself with a streetful of beauties. It possesses today the longest, most complete, most systematic histories of wild primates to be found anywhere on earth.

The Japanese monkey *Macaca fuscata* is closely related to India's rhesus. Like the rhesus, it lives in large, highly organized troops, spends most of its time on the ground, and adapts itself to a variety of environments from Japan's cold, snowy north to the sub-tropical south. Unlike the rhesus, it has a pink face and a stubby tail. Systematic study might be said to have begun on a warm little islet called Koshima, when in the summer of 1952 two scientists, Junichiro Itani and Masao Kawai, found that they could induce a troop of seventy-odd monkeys to eat sweet potatoes. While in all other ways the group retained its wild patterns, it reported regularly for the sweet-potato ration and so individuals became familiar and relationships subject to study.

"Provisioning" became the basis for Japanese technique. The following year the two Kyoto University scientists introduced provisioning to the largest troop of monkeys in Japan. They lived on a forest-clothed mountain, Takasakiyama, above the

Pacific, and came down to the feeding station near a temple group. By the time Carpenter visited Takasakiyama in 1966, visitors paying admission to watch the show numbered 150,000 a year. Yet the monkeys remained wild. And by then Japanese science had accumulated not only the case histories of hundreds of individuals, but family trees as well. By the single artifice of provisioning, the Japanese made possible the study of individuality in otherwise natural groups. Thus a record of revolution was made possible.

I have mentioned how infrequently animal social orders are upset by revolution from below. And perhaps the Takasakiyama revolution of 1959 was in part a consequence of provisioning and undue population growth. By that time the colony had reached an improbable primate number of 700. Its structure was orderly, grouped in two concentric circles. In the center were six high alpha males, each with subtle rank but, like the baboon oligarchy, cooperative, acting in concert and without quarrels. With them were the females and infants, along with older but immature females. One portion of alpha responsibility was breaking up quarrels among them, the other that of keeping young males out of the inner circle. The outer circle, their province, was dominated by ten younger sub-leaders, each, like the high alphas, with subtle individual rank. The elaborate structure kept social order, and reduced fighting to a minimum. But by 1958 Itani was reporting growing stress. The next year the explosion occurred.

The highest central alpha was called Jupiter. And one day he was attacked by a high young peripheral alpha named Hoshi. It was the success of the outrageous attack, one must assume, that demoralized the social order, and the next day Hoshi and 250 followers were gone to establish a new troop. Now the latent territoriality of the Japanese monkey was demonstrated. Previously there had been the one colony. With two in the area, a frontier was established, and border war for a while was continual. Hoshi, we might say, had sufficient charisma to attract the following response of over a third of the group. But old Jupiter, despite his defeat, retained the loyalty of most of the original group until his death a few years later. Shortly thereafter the still growing group split again, so that now there are three troops on the green slopes of Takasakiyama, each with its exclusive territory. When Carpenter made his visit, however, relations had settled down. Territorial belligerence had been reduced for the most part to ritual. An aggressive male would climb a "shaking-tree," there to rattle its branches in a clenched-fist gesture of warning and defiance.

It has been in the area of social learning that the Japanese have made their most spectacular contributions. I have suggested that learning, a prime function of society, makes the superior capacity of an individual available to all. When sweet-potato feeding was begun on the little islet of Koshima, there was a female only eighteen months old who bore no visible sign of formidable IQ. Not too far from where the sweet potatoes were normally dropped on the beach, a little stream crossed the sand into the sea. And quite shortly the girl monkey was taking her sweet potatoes to the stream and carefully washing them of sand before eating. By now Syunzo Kawamura was the observer. And before long a male of her own age was doing it too. Then her mother learned. Slowly the new cultural institution spread, and Kawamura kept his records. But the invention never reached those older males who had no attachment to the young.

This remarkable observation naturally gave the scientists ideas. It occurred to Itani to introduce a new food to his Taka-sakiyama colony, then still a single troop. He chose caramels. Slowly the strange food caught on, spreading just as the potato-washing had spread. The young were the experimenters, mothers and other young followed, and new infants learned from their sophisticated mothers. Again, no adult or sub-adult lacking contact with young ever acquired a taste for caramels. And in the meantime Kawamura was trying another novel food, wheat, with startling consequences.

The Japanese were beginning to believe that they had found among the young the normal course for the spreading of a cul-

tural innovation. Kawamura was from Osaka University, and
near Osaka is a ravine called Minoo. Here two troops of monkeys had their traditional territories; Kawamura chose the
second of the troops for the wheat experiment. And in
the Minoo-B troop all conclusions were challenged. Here the
highest-ranking male—the supreme alpha fish—immediately
started eating wheat, followed promptly by the alpha female.
And whereas Itani's caramels had taken a year and a half to
spread to half of the Takasakiyama troop, in Minoo-B all were
eating wheat in just four hours.

Wheat for the original Kashima potato-washers followed the
normal course, but it too produced a stunning event. The wheat
was scattered on the beach, and the monkeys patiently picked
out grains from the sand. But the same little female genius, Imo,
who at eighteen months had started the sweet-potato washing
and was now four years old had small patience for the process.
She scooped up her wheat, sand and all, and, applying the best
practices of placer-mining, went to the stream, where she let the
current wash the sand away between her fingers. When Car-

penter visited the island in 1966, she was an old monkey, but she was still washing sand from her wheat, as was many another who had learned from her. And perhaps new innovation was in process. Young monkeys were going downstream to catch those floating grains that escaped her fingers.

That the lowly Japanese monkey makes the lauded chimpanzee seem a dolt has perhaps its rationale. Once in conversation in London, Kenneth Oakley made a suggestion which I have never seen developed in the literature. Oakley is keeper of anthropology at the British Museum, and with Bernard Campbell of Cambridge is our foremost authority on fossil man. We were discussing the critical importance which physical anthropologists attach to brain size in relation to human evolution, and Oakley was discounting its significance. In social animals, as he saw it, intelligence must be measured in terms of social organization and communication. If Neanderthal man—frequently with a brain larger than his successor, ourselves—failed and became extinct, then one should look to defective social organization for an answer. Brain-power represents, in other words, not just the effective organization of your nine billion cells (if you are a human being) but the effective flow of cerebral resource within that community of which you are a part. If that community is divided, heedless of organization, incapable of communication, then the power of the individual brain can no longer be measured in terms of neurological resource.

The success of the Japanese monkey lies not in his unremarkable brain but in his remarkable talent for society. He submits himself to authority, but is capable of revolution when things get too thick. Despite the rigidity of hierarchy, still equality of vertebrate opportunity is such that a superior but immature female—about as close as one can get to the ultimate omega—can make her contribution to the total brain power of present and future Japanese monkeys. That the flow of learning is incomplete and does not reach unconcerned adults is of no significance if we think in terms of future generations.

What is of significance to any present generation, and what must be regarded as the supreme discovery of the Japanese scientists so far, is the spectacular spread of social learning when innovation and alphaness coincide. The omegas of Takasakiyama took eighteen months to spread a taste for caramels to 51 percent of the troop; the alphas of Minoo-B took four hours to establish a taste for wheat *in toto*. If we continued to think of alphaness in terms of strength, might, despotism, then the consequence would be inexplicable. But if we see it in terms of Chance's "attention structure," or of the following response, then the value of the alpha to every social member becomes apparent. The omega commands the attention only of his or her intimates. The true alpha holds the attention of the entire group, and through giant magnetism accomplishes as a normal event what would otherwise be a miracle.

We need not labor long in the human vineyard to gather our tasteful or tasteless fruit. The power of fashion demands no comment other than that man did not think it up. The dazzling gift to a wartime Britain of a Winston Churchill, the sickening tragedy in a peacetime America of a John F. Kennedy, speak of the value of the alpha to every last omega. High alphaness is rare, and we know it when such alphas are gone. But Adolf Hitler was an undoubted alpha.

Neither reader nor author is yet in a position to judge facts of social life when we have only begun to accumulate them. I myself am grateful indeed to the scientist in a ravine just outside Osaka who made a contribution of such significance; but I do not even know his first name. Kawamura placed him in charge of the Minoo-B experiment, and his name was Yamada, and his paper has never, to my knowledge, been translated from the Japanese. One may read about it, however, in Kawamura's contribution to Charles H. Southwick's valuable and available *Primate Social Behavior*. And one may reflect that it is out of such anonymous dedications on far distant fields that the sciences are bringing us evidences of man.

5

In the spring of 1967 I was starting a tour of lectures at Pennsylvania State University, and Carpenter and I had the pleasure
and peace of the Easter weekend at his spreading house in the
nearby woodlands. We were chin-deep in the films and the
photographs that he had taken of Japanese monkeys the previous summer, but somehow we turned to speculation concerning
the life of a monkey in a society of 700. And the psychologist
Carpenter made a memorable analysis of monkey decision-
making.

"You are a monkey," he said, "and you're running along a
path past a rock and unexpectedly meet face to face another
animal. Now, before you know whether to attack it, to flee it, or
to ignore it, you must make a series of decisions. Is it monkey or
non-monkey? If non-monkey, is it pro-monkey or anti-monkey?
If monkey, is it male or female? If female, is she interested? If
male, is it adult or juvenile? If adult, is it of my group or some
other? If it is of my group, then what is its rank, above or below
me? You have about one fifth of a second to make all these decisions, or you could be attacked."

I have never encountered a more eloquent, more realistic, or
more humbling tribute to the capacities of the animal mind. In
any social group that includes rank order as its structure, every
animal must know every other animal individually, and be aware
of himself at the focus of the varying relationships. Thus,
through social ranking, natural selection places a premium on
intelligence, for the animal too stupid to meet the competition
of Carpenter's decision-making process will probably be eliminated as a breeding factor in the group.

To the successful organization of unequal beings in confrontation with common needs, hierarchy makes many a contribution. It reduces fighting. Once the order of dominance is
established, serious aggression becomes rare, since each member

knows too well his own capacities in relation to the next fellow's. Rank order sorts through competition the unequals, placing in positions of influence those with superior assets for the group as a whole. Then, through social learning and the following response, it makes their achievements available to become individual assets of every last member. There are genetic consequences as well, since in most species the sexual attractiveness of high rank and unattractiveness of low favors a disproportionate contribution to the gene pool on the part of the highly endowed. Hierarchy even disposes, in the strangest of ways, an animal justice, and we shall look at that in a moment. But perhaps the most significant contribution to the species as a whole, and to evolution in its long upward sweep, has been the demand placed by status on intelligence.

When Konrad Lorenz bred his first flock of jackdaws—this was back in 1927 at his old home in Austria—one of his first impressions was how young they are when each recognizes all others individually. Jackdaws pair in the first springtime after they are hatched, and do not mate for another year. But long even before the pairing, the young males have completed their shuffling, their pecking, their crowding, their staring each other down, and have established the recognized rank which they will probably occupy for the rest of their lives. The females recognize it too, for with the pairing each assumes the rank of her male, and knows which other females she must defer to and which she may push about. And while Lorenz' first flock was small, only fourteen birds, what he found has been confirmed a thousand times over and in far larger groups.

Noble wrote of social rank's uncanny eye for detail. And such must be the animal eye that so quickly assembles its social register. Dairy cows have their butting orders, and will follow them with 95 percent consistency. The sea otter would seem to challenge the eye for detail, since most of the animal is under water. But when a herd of one hundred was carefully observed for a season off the California coast, a remarkable order was revealed. Otters—sea-going baboons in the sense of their talents for banditry—steal food from one another. But each knows whom he can steal from, whom he cannot. Sea-otter rank may be described as no-steal-back.

Glen McBride, of the University of Queensland, has made many observations of chickens valuable to Australian poultry-growers, no doubt, but as valuable to us. Any society based on rank order, he concluded, must have a maximum number reflecting the animal's capacity to discriminate between individuals. With hens the number is about fifty. The flock that is increased in size will suffer disorder and increased aggression through declining social control. Wild species may differ in this particular evidence of animal IQ, and other factors such as food supply may limit the social number. But a final limitation must be one of intelligence. Thus, baboon troops in East Africa average forty or fifty in number. But in Mozambique's Gorongosa reserve, with its abundant food supply, I have seen thoroughly integrated troops running over two hundred. Baboon intelligence is equal to such identification.

In a way, McBride's concept of the limits of individual identification supports Altmann's definition of the social border as that line where communication ceases to be effective. Altmann's is a broader concept since it will apply to any group, such as the starling flock, that responds to the same alarm signal, whereas McBride's refers only to groups with a structure of organized social order. But the problem remains that of communication. The alpha can release the group's following response only so far as members can identify with him as individuals. And he can exercise his responsibilities only so far as he can identify his subordinates.

Alpha responsibility is difficult to comprehend, for in many species it is as if the alpha, in exchange for his privileges, assumes definite obligations. Again let us return to Lorenz, since he was the first to make broad observations of social behavior. When he had bred his second flock of jackdaws, he found himself with a large mixed flock of older and younger birds. One day a huge flock of migrating jackdaws and rooks settled in an adjoining meadow, and all of his younger birds flew immediately to join it. His consternation was painful, for it seemed to him certain that when the big flock moved on, his young ones would follow, and he had no way to retrieve them. But he had not given proper estimate to the responsibility or the resources of his high-ranking older birds. They had stayed aloof from the excitements of the visitors. Now they flew out, selected one by one their young fellows from the hundreds in the meadow (Lorenz' eye could not identify them), and by swooping low induced a following response and led them home. Lorenz lost only two birds.

Defense of the group is an alpha commonplace. I have mentioned the gang at the top of the baboon troop that with any alarm signal goes immediately to meet the enemy. In a famous experiment with a captive group of rhesus monkeys at Calcutta, Southwick introduced every variety of stranger. All were attacked, but when juveniles were introduced, 70 percent of the attacks were by juveniles. The alpha remained aloof. When females were introduced, almost all the attacks were made by females; two strange introduced males were attacked both by the beta male and the females. Through all, the alpha avoided the commotion. But when a human handler entered to net a monkey, he was instantly and viciously attacked by the alpha. Southwick concluded that it was external threat to the group that roused the leader.

Hall's observations of the patas monkey in Uganda suggested that the alpha's function was almost entirely that of sentinel and guard. The patas male is large, so conspicuously colored that he is sometimes called the hussar monkey, and, living a terrestrial life on the savannas, is the fastest-running of all primates. He lives in a one-male group with half a dozen or more drab females and their young. So intent is he at scanning the savanna for danger that actual movements of the group are usually led by a female. But if danger appears, his defense is remarkable. Once Hall surprised a group. To his own surprise, the male at blinding speed ran directly at him so that Hall thought he was being attacked. At the last moment the male veered off, while the human intruder, still in shock, looked after him. Then Hall looked about for females. While the male had conducted his diversion, they had vanished into the savanna.

If in a prairie-dog town in the American plains a subordinate gets into an argument with a member of an adjoining clan, the alpha will drive him away and take over the fight. When Scottish red deer are disturbed, any may give the first alert, but after that only the alpha hind. And when the party withdraws she will be the rear guard. I have watched the same action in South Africa's Kruger Park when, coming late into camp, I was halted by a party of frightened kudu breaking across the road and vanishing into the dusky bush. Like red deer, the bulls of this superb antelope species live off on their own while mothers and young form their own party. For moments after the cows and their young had disappeared in the bush, a last female—one may safely presume, the alpha—remained at the road's edge, peering back into the dark for the possible lion while she snorted warnings to her now-distant company.

Social defense is not always the obligation of the alphas, but, as in the Japanese monkey, may involve a fighting class of younger peripheral males. We cannot know: they may be merely redirecting their aggressions. But what I have referred to as animal justice is invariably the duty of high-ranking males. And its administration is surprisingly simple.

There are times when anthropomorphisms are almost irresistible to even the most rigorous of scientists. As unlikely victims as Niko Tinbergen and David Lack have found the word "righteous" inescapable when describing the behavior of a territorial proprietor threatened by an intruder. So the adjective "aristocratic" is most difficult to avoid if students are to give any clear impression of the alpha. He holds himself aloof, as did Southwick's rhesus. He may bicker with his beta, but, like Lorenz' alpha jackdaw, he will avoid discussion with all those farther down the rank. He will bicker with nobody if he belongs to the baboon oligarchy. The gorilla silverback, despite his mo-

nopoly of power, will look on with indifference while a subordinate copulates nearby. So will the alpha vervet and the dominant bull elk; the male lion will go to sleep while his brother-in-alphaness handles the situation. The rhesus or Japanese alpha male moves like royalty while deference clears a space about him. Altmann and Chance measured that space in both the wild and captive rhesus and found it the same, eight feet.

The aristocrat stays apart from hoi-polloi and avoids all unseemly actions. But as in most species he has a responsibility for defense, so in many he has an obligation for order. If a fight breaks out among the rank-and-file, he will leap in to break it up. And, breaking it up, he sides inevitably with the weaker. So must it be with the rooster bringing peace to his quarreling hens, or the ugliest of baboons putting a juvenile bully in his place.

Perhaps there is no mystery about animal justice, strange though the phrase may sound. Presidents and kings appeal to the people in their struggles with the barons. Fathers must defend their most intolerable children if the family is to be saved. When order is essential to the survival of beings, then alpha and omega must make compromises, however distasteful; concede sacrifices, however repugnant; make communication, however boring; accept risks, however dangerous; seal alliances, however temporary. Animals, in part because instinct allows no real choice, in part because intelligence is unclouded by false instruction, succeed. But man—suspended between dicta three billion years old and a foresight *nouveau riche*, swinging between wisdoms of most ancient origin and a power of both learning and ignorance—here is the animal of doubtful future.

5. Time and the Young Baboon

In the later decades of the twentieth century the revolt of the young absorbed the debates of adults as passionately as had in previous eras the assaults of Moslems on Christians, the bloody disagreements of Protestants and Catholics, the horror of aristocrats facing the bourgeoisie, the terror of the bourgeoisie facing the proletariat, the shock of imperial masters facing the claims of colonials of whatever color, the pallor of auction-bridge addicts watching the encroachments of contract, and the implacable fury of devotees of New England clam chowder at the grossness of taste among those who would accept Manhattan clam chowder. There was a difference, however. Earlier clashes had been temporal, enduring for no more than a few centuries. But the revolt of the young seems to some—and I am among them—quite possibly a permanent feature of society's future landscape.

An evolutionary review of the relations between the young and their elders may do little to relieve our temporary agitation, but may do much to provide longer enlightenment. And a good place to start is with the work of a man named Gene Sackett, conducting experiments at Harlow's Wisconsin laboratory. His study was published in 1966 in widely read *Science*, and I must assume that skeptics have had time and opportunity to register their skepticism. This they have not done. Yet the study revealed an innate line of division and antagonism between young and old.

Sackett's experiment skated on thin laboratory ice since he raised young rhesus monkeys from birth in total isolation, and we have seen what neurotics may be produced. Harlow himself, however, had demonstrated that a mere twenty minutes a day of contact with peers was enough to ensure normal development. Sackett's ingenuity was to substitute for such contact a translucent screen for one wall of the cage, and rear-projected slides for social stimulus. And it worked. None of his subjects became the old-fashioned neurotics clutching their heads, biting their skins. The daily visual stimulus was enough to contain monkey sanity, as the shadows on cave walls were enough to keep Plato busy.

The experiment began when infants were fourteen days old and had their first exposure to the cave wall. It extended for months. Their experience included nothing but the shifting images of the slides. Ten categories were provided, with so many different subjects in each that none was repeated in a week's time. Nine categories were of monkeys: such situations as mother and infant, adults copulating, an adult threatening, infants playing, postures of submission or fear. A tenth category was of non-monkeys, whether pretty girls or early American furniture. A certain level of excitement was induced by the very appearance of an image on the screen, but the response to non-monkeys was notably at a minimum from first to last. And although the question was not the subject of the experiment, still one must wonder: How did these monkeys know they were monkeys?

The varying response of naïve monkeys to the different categories of monkey deportment was of course the object of Sackett's experiment. And all categories elicited degrees of vocalization, activity, stimulation, even investigating the screen. But with the exception of two categories there was no great variation in response. All subjects were interesting, so long as they presented monkeys. But I doubt very much that the observer anticipated which two monkey situations would prove the exceptions, galvanizing the attention of all: pictures of infants playing, and the sight of a threatening adult male.

Men and apes threaten with a frown, but, like many monkeys, the rhesus threatens by lifting the eyelids in a stare that exposes the eyelid skin. Anyone would assume that adult threat is learned through hard experience. But it was not so. The sight of a threatening adult brought cries, cowering, huddling, wall-climbing, resembling the response to no other situation. The adult threat is a stimulus in the naïve rhesus releasing an innate, terrified response. The peak of the disturbance is reached at the age of about three months, then gradually drops away. Since nothing too terrible had happened in consequence, presumably the young monkeys became used to threat.

Response to the sight of infants playing was as decisive as, but more lasting than, response to threat. As I have asked how did these isolates know they were monkeys, I may ask how did they know they were young monkeys? Yet just as inborn were the responses of excitement, scampering, calls, investigating the screen, responses of undoubted attraction for their peers. The reaction to a mother and infant or to adults copulating received the normal ration of interest greeting any stimulation. Only play with one's peers and fear of one's elders could be described as innate in the infant rhesus.

As we consider our human young and the widening gap between generations, we shall do well to inspect the alpha-omega relations of other species. We need not confine ourselves to such primates as the rhesus, or to laboratory experiments. In almost every society of animals the line of maximum tension lies between the maturing males and the adults of the male establishment. Variations in this line—seldom pleasant from the viewpoint of the young—rest largely on the dispensability of males, and sometimes, when birthrate is high enough, on the dispensability of everybody.

Beaver kits at the age of two are thrown out, regardless of sex. There is a very good reason, though the two-year-olds may prove lacking in understanding. A beaver pair makes a huge capital investment of energy in the building of a dam, the conversion of some Canadian stream into a fair-sized pond, and the construction of lodges. They establish a defended territory of woodland around the pond to assure a supply of saplings for food and dam-maintenance, but it cannot be large or dragging back saplings from a distance will bankrupt the family's energy. The population is therefore limited. The parents do not hesitate to exploit their young, and so until the age of two the young beavers, being of a hard-working line and likewise unaware of their fate, toil with a will at the cutting and pulling and stripping of saplings. Then the new batch of kits is born and out the two-year-olds go. They will wander, encountering foxes here, famine there. Lucky ones may find a good spot for a dam, and pair up, and start things new.

A chipmunk mother drives her young off her territory two weeks after they leave the nest. When W. H. Burt was the great man of Michigan's zoology department, he meditated on the lives of his favorite little beings haunting the meadows and woodlands near Ann Arbor. Wood mice, pine voles, white-footed mice, short-tailed shrews—all faced the most hazardous moments of life when driven away, still young, from their familiar terrain, their assured hiding places. Few, he wrote, would die of old age.

So goes it with the young of species in which mathematics, not mercy, will provide sufficient survivors. In most species, however, a bias prevails in favor of saving the females. Whether human devotion to saving women and children first springs from courtesy or conservation, I do not know. In many species it is conservation of the child-factory. A dimmer view is taken of the male.

Male lions are driven out of the pride at three. Since the lionesses do almost all of the hunting, the male becomes little but an ornament with vast appetite. It is true, of course, that the ornamental value cannot be discounted, and one or two great long-maned beasts should be kept around to thrill tourists. It is true also that the sexual appetite of the lioness is such that males cannot be entirely counted out. George Schaller's study of the mountain gorilla broke some kind of ethological record; in his recent three-year study of the lion in Tanzania's Serengeti he observed the sexual capacity of a lioness that must break some other kind of record.

There is a huge two-male pride which, since its territory is not far from Seronera Lodge, is familiar to many visitors to the Serengeti. A lioness came into heat and Schaller enlisted for the duration. The two alpha males reported likewise for duty. One got there first, the other went to sleep. In two and one half days the lioness and the wakeful male copulated 170 times, at a regular rate, around the clock, of about once every twenty minutes. After Number One Hundred and Seventy, the male, for reasons scarcely obscure, disappeared into the bush. His partner woke up and, since the lioness was obviously not yet appeased, took over.

Alpha capacities in the lion establishment are perhaps as discouraging to the maturing young lion as are alpha incomes to the maturing human student. I do not know whether the three-year-old lion takes off because of alienation, of disgust for his elders, of an Adlerian problem of inferiority, or simply because he is kicked out. But he will become a nomad. And as he becomes older, and heavier, and more slow-footed, his chances of survival are low unless he finds females who will be enchanted by his alphaness, and will support him.

The maturing male of whatever species does not have it good. The sexually maturing male elephant must leave his mother and join the male band, with rank order already established, where he will find himself omega. "Psychological castration," a phrase invented by the American zoologist A. M. Guhl, is a common event in some species. The alpha buffalo assumes all sexual prerogatives; his male companions remain indifferent. The young male Uganda kob joins the all-male herd; if he secures high enough status, he will enjoy the privileges of the stamping ground and its sexual opportunities. But of any 500 male Uganda kob, only fifteen or so will at any one time be found on the stamping ground. The remainder seem to enjoy their lives, but they are psychologically castrated. So it is with almost all antelope species. A final indignity is placed on the omegas by territorial wildebeest males, who take all the best pasturage. Here the females and young of the nursery herds may graze, while the non-productive males are consigned to the stubble. One may observe a grand design in evolution's injustices.

Such a grand design seems to have impressed Bristol University's astute ethologist John Hurrell Crook, whose observations of weaverbirds I have reported. He went on to study the hamadryas baboon in Ethiopia, and that halfway house between the baboon and other monkeys, the gelada. Both, like Hall's patas monkey, have the one-male society with a harem. He watched his geladas—big brown animals with manes like the hamadryas—as family parties moved along feeding together while peripheral, psychologically castrated males were, like the wildebeest, condemned to the poorest tables. And it occurred to him, and to J. S. Gartlan, his colleague at Bristol, that in all the primate world there are known only three species in which a male normally monopolizes a harem. And these three live in the huge African area grading from Uganda's dry northern reaches through Ethiopia to the Sudan and, presumably, at one time to Egypt. Monkeys and apes live usually, like affluent men, only in verdant suburbs. But these three species, having to make their peace with arid environments short on food, all took a common road in the evolution of their societies. The solution was simple: dispense with unnecessary males.

And so we have the polygamous households of the primate in arid lands. An alpha male takes over as many females as he can attract, keep pregnant, protect, and put up with. For every extra wife, a male becomes redundant. An outcast, he vanishes into a marginal existence of psychological castration and undernourishment. In such species the establishment of alpha males offers few openings for any but the most extraordinary young.

Perhaps one cannot say that in the species I have been describing a visible line of tension exists between the mature and the maturing. The line is erased by banishing young males, or all that you do not have jobs for. However appealing the pattern may be, it is a solution unavailable to *Homo sapiens*. If you belong to a species in which you need your strong, healthy young males to hunt your food, defend your borders, conquer your neighbors, or stand as sentinel through nights where leopards prowl, then such simple solutions will not do. An accommodation, however disagreeable, must be found between the establishment and the candidates, but it is a rare species that accomplishes it without the line of tension. The problem remains always the same.

All juveniles are omegas, since all adults are dominant over all young. In this sense, the problem of the adolescent male is his omeganess. But there is a sea of anguish dividing the psychic shores of the adult omega and the adolescent omega. In any animal society the adult male who has settled to a lowly rank has done so through some deficiency in those qualities that make for alphaness. Whatever be his deficiencies, whether of strength or vitality, of courage or determination, it is unlikely that he has ambitions beyond existence. He makes his peace. But the adolescent bears only the low rank of youth. It is a deserved omeganess, since if his society is to survive, then he must have reached a fullness of development guaranteeing his responsibility when he enters the ruling ranks. It is an omeganess too that protects him, while he matures, from participation in the status struggles of the adult. But it is a class omeganess, and though his fires burn hotter, his muscles grow stronger, his intelligence grows keener, his ambitions fly higher, he must bear his stigmata until that day when maturity flings him into the establishment, an intruder to be resented and resisted by all.

The young do not have it good. Youth, shielded by its omeganess, may enjoy its careless raptures. But a condition of being young is time, and there is the fatal truth that being young cannot last forever. And if you are a maturing male in a species wherein young males are indispensable, then as your moment of truth approaches, you will have but one consolation—that it will bring everyone just as much trouble as it brings you. I know of only one species where accommodation on the part of adults has been so spectacular as to eliminate all pain. But that is a

species in which the indispensability of young adults becomes in itself of tragic proportions.

The African hunting dog, known in the early days as the Cape hunting dog, is perhaps the world's most successful predator, excluding only man. It is not a dog at all, not even a member of the family *Canidae*, but has four toes on his foot, ears like ping-pong paddles, and bears the scientific name *Lycaon pictus*. It looks like a dog, however, and not a very large one, since it weighs about forty pounds. Its success as a hunter has been based entirely on a capacity to outrun any animal in Africa, together with the superb coordination of its hunting pack. But despite its success, or because of it, the hunting dog is a rare animal today. Since the early times of settlement in the Cape, three hundred years ago, its reputation has been so terrifying that men have hunted it, even poisoned it, as vermin.

The hunting dog has terrified not only men but other animals. I have watched Thomson's gazelle graze within two hundred yards of a hungry cheetah, the fastest of land animals. But in 1966, when Eliot Elisofon and I had the chance to photograph a large hunting-dog pack for three days, dawn to dusk, no antelope approached close enough to be identified without binoculars. Within our view was perhaps thirty square miles of flat, open, treeless plain; along the margins animals grazed. Just once four zebras strolled by at a distance of several hundred yards. Since they had the confidence, one must assume that the margin was sufficient for escape. But also the pack was sleeping.

At that time a study of one pack, over a period of a few months, was all that science had yet collected in the way of reliable observation. When I returned to the Serengeti two years later, however, Kruuk and Schaller, as by-products of their hyena and lion specialties, had collected more information on the entire predator community than science had ever possessed before. It is information which, although largely unpublished, I am permitted to use before this investigation closes. Now, in terms of adult-young relations of almost shocking amiability, I wish simply to describe one hunt.

Schaller, by the summer of 1968, had followed twenty-two packs. He came by my cottage about four o'clock one afternoon, having spotted a pack which he judged would hunt about five thirty. By four thirty we had found it, sleeping in the midst of a rolling area of plain parched by the dry season, burned black by fire. We animal-watched, waited, talked. There were ten adults and fourteen pups well over half grown. Schaller had known the pack when the pups were born six months earlier and there had been sixteen. They had lost only two. It was an infant mortality rate so low as to be difficult to believe. Finally three adults were stirring, nosing the others to their feet. All went to the pups, and now there was a rolling pell-mell of play, adults and pups together, and the strange twittering sound, like a flock of sparrows, that the dogs make when excited.

"War dance," said Schaller. And like a ritual it was, for the dogs when sufficiently excited were all on their feet. The three leaders headed off, other adults behind them, then the pups, and last a disabled adult on three legs. It was precisely five thirty; whether they had heard a factory whistle blow or Schaller earlier had given them instructions, I do not know. One of the most memorable sights in nature, however, is that of a pack strung out in single file almost a quarter of a mile long, headed into the late, gray light with white-tipped tails upraised like beacons to make following easier.

They trotted. The hunting dog—like the albatross, the elephant seal, and few other species—has no least fear of men, and one may drive beside the pack as if one were not there. We checked the Land Rover's speedometer, and they were doing fifteen miles an hour. The pups kept the pace with ease, as did the adult on three legs. Now the leaders speeded up to a run. The speedometer read twenty-five miles an hour. Still the pups and the three-legged rear guard kept even. But now the leaders were putting on pressure and a gap was widening between adults and pups.

Since game flee an area in which a pack is running, a common tactic is to approach a rise at high speed on the chance of surprising prey on the far side. Such speed for the three leaders, with whom none could keep up, was over forty miles an hour. They outdistanced us, and by the time we reached the rise the three had stopped, the others were catching up, and across the empty valley beyond, Thomson's gazelles were ototting half a mile away. The word was being spread. Hunting dogs kill by surprise in ten or fifteen minutes or have a long search ahead through a forewarned world. Yet 85 percent of all hunts observed by Schaller have been successful, a record to stun a lion pride.

The leaders were moving again in a new direction at a modest run, the whole torchlight parade of bouncing white tails reassembled behind them. Then the leaders were digging in at full run and we could see three large wart hogs ahead running at an angle to our left toward their hole. That monstrous machine, the Land Rover, will survive cross-country driving at such speed, but only with difficulty will its human occupants. I lost the wart hogs. Within my view, however, the dogs were beginning one of their notable maneuvers. When prey curves in its flight, each dog in the file sets a separate course so that the pack is like narrowly set spokes in a wheel. They are not following; that is left to the leaders. They are setting courses of interception along the probable curve. But now there was no necessity. The wart hogs had vanished and we thought they had made it underground. But the leaders were struggling, for they had caught the last one by the hind legs and all one could hear was the squealing of the wart hog.

"Poor pig," said Schaller. Like a first-class nightmare, one is unlikely to forget a hunting-dog kill, since it does not kill but eats its prey alive. Nothing has so contributed to hunting-dog horror as the long-drawn manner of its prey's death. But in truth the dog cannot kill. He is small, and his teeth are small, even for his size. He has forty-two, and they are designed for slicing, not killing. Only the cats have killing teeth; even the wolf, taking caribou or moose, will proceed as does the hunting dog. And yet, while nature may provide its explanations, the spectacle remains other than nice.

I find it a subject for meditation that within my limited career of animal-watching, the two images most horrid and sublime were separated by little over sixty seconds. By now the hundred-pound wart hog's viscera were torn open and he was finished, though still squealing. Then the fourteen pups arrived, crowding into the living feast. And with their arrival every adult stepped back, none with more than a mouthful of meat. They were undoubtedly hungry, for their bellies were lean. Now and again they would string themselves out in hunting formation at some hint of further prey. But in the hour and fifteen minutes that Schaller and I sat beside the kill, no adult took another mouthful of meat. The food was exclusively for the pups.

We sat in the gathering darkness in the middle of a plain emptied of all living creatures by fear of these formidable little beasts. And we discussed what course of natural selection could have produced such inhibition favoring the young. Only a very high adult death rate could place such selective value on the successful raising of young to maturity. I speculated on disability. In the lion pride all cubs eat last, and if kills are few or small, they starve. But the adult lion is very nearly invulnerable. The hunting dog, small and relatively fragile, may suffer a high casualty rate when attacking prey more formidable than wart hogs. Schaller shook his head. "I think disease," he said.

When it grew too dark to stay longer, we drove away. And six months later Schaller wrote me that distemper had hit the pack, leaving only nine survivors. When the young have it good, there is a reason somewhere.

2

The age group is one of the commonest features of animal social life, and in our changing times it is among the most poorly understood phenomena of human society. When you read Frank Fraser Darling's A *Herd of Red Deer* and find companies of stags—mature males—in which those of an age tend, despite maturity, to keep together; or when you hear that in an immense school of mackerel those caught within a given area tend to be of the same size: then you may begin to wonder about your contemporaries.

There are the young of the same age in an Israeli kibbutz who from infancy play as a group, go to school as a group, enter the army and fight as a group, and for whom members of a peer group throughout all their lives will probably find deeper attachment than any other association they will ever make. It is the same in the age groups of many an African tribe. Something of the age group is found in the mobilization of armies by year of birth, or the nostalgias of the American college graduate for the class of '52. We who are of an age will always understand each other a degree better than we understand others.

No human arrangement could more easily be dismissed as a product of conditioning or cultural institution. But we must recall Sackett and his infant rhesus monkeys. If an innate demand for one's peers exists in a primate species as highly evolved, as socially adept, as capable of solving complex laboratory problem as the rhesus, shall we dismiss remnants of such demand in men? Frequently a human cultural pattern exists to implement an underlying biological pattern. And I suggest that we can begin to comprehend the irrationality and frequent absurdity of the student revolt beginning in the 1960's only if we recognize its ancient, irrational foundations.

A line of tension between mature and maturing exists in most animal societies. Its visibility is increasing in human societies today. If a communication gap of abysmal depth widens between mature and maturing, and if at the same time we behold arraying against the elders an alliance of the young within which communication seems perfect and mutual perception almost of an extrasensory sort, we may well deplore it as a grenade in our social bowels that could destroy us all. But we are permitted to wonder only so long as we are permitted those comforting illusions of human uniqueness and rational sovereignty, both of which have contributed so splendidly to the area of social disaster in which we serve.

The thoughtful, disturbed adult may say, "But why now? It wasn't this way when I was young." And I must agree that it was not this way at Chicago in the Class of 1930. But we have all of us been seduced. The Freudian enthronement of sexuality as life's prime mover has reached its climax of acceptance— by ourselves as much as the young—just as ethology begins to prove that it is not normally so. And the second Freudian seduction was that the family for all eternity is the unit of human affairs. Within its sex-empurpled embrace the attachments and antagonisms of early childhood become the determining agents of life. Jung was wiser. He dealt in broader contexts which brought the crisis of adolescence and the conflict of fantasy and reality into a focus of greater relevance for our times. We shall do well to re-study Jung as we re-study Adler.

What we watch today is the disintegration of the family which we were taught was universal and eternal. It is neither. The family was never the windowless chamber as the Freudians saw it except in isolated, insecure arrangements of *Homo sapiens*. Neither has it been eternal. Its significance was sparse, I suspect, before farming succeeded the communal hunting band ten thousand years ago. But if the technology of farming did or did not create the individual family as a significant social unit, then with small argument it is technology, man's Frankenstein, that is destroying it.

The adult who asks "Why now?" must recall that only since the Second World War have technological advances, together with the vast organization of technological empires, reduced the family as a unit to microscopic scale. The young have discovered its insignificance; you and I have made it so, but we have not looked. And while we have been seduced by our view of the family as imperishable, we have been deluded as well by economic determinism. Capitalist and commissar, in perfect agreement, accept material satisfaction as the human rainbow's end. And while the most apoplectic defender of free enterprise in truth confesses in the Marxist box, youth turns away. Affluence after ten thousand years catches up with us. The family, that final cooperative unit in the war against want, disintegrates for lack of function. And the stricken father sighs, "I gave him everything I could."

A virtue in the study of primate societies is that we study social organizations not only recalling the human past but perhaps anticipating the human future. We may put aside for the moment the problem of affluence, and the question of why a son raised in luxury and facing a future of material security has vanished into pot or protest. The question can wait. What concerns us is the power of the age group as it replaces the power of the family.

A significance of the primate is that his societies are seldom organized around family units, and unless nuclear lightning strikes us down, restoring to the few survivors the deprived conditions of our pioneer past, this seems the likely human future. In 1961 I wrote that the family is the building block of primate social life; it was an informed guess based on the scanty evidence of the time, and the information ran out. I was wrong. In none of the larger primate societies—those such as the olive and chacma baboons, or the rhesus and Japanese monkeys—is there a vestige of the family unit. And the evidence suggests that never in the evolutionary past has this been the principal primate way.

The true lemur is a pro-simian—a pre-monkey primate—dating back fifty million years to the Eocene before monkeys evolved. In his studies of the black and ring-tailed lemurs lingering today on Madagascar, Jean-Jacques Petter found mothers driving off their young at the age of six months, whereupon the young promptly formed age groups. Of the two great monkey sub-orders, the Old World and the New, the American species seem the more primitive. In his early studies of the howler and red spider monkeys in Panama, C. R. Carpenter found no trace of family life. The bond between mother and infant yields shortly after a sibling is born, and the older moves off into the world of its peers. The Old World species I have mentioned—the baboons and rhesus and Japanese monkeys—and others like the African vervet or the Asian langur follow the same pattern: the age group, not the family, is the juvenile's home. We may say with cynicism that technological man is returning to the way of the monkey. Or we may say, with greater reward, that to understand ourselves we have the fortune of the monkey to inspect.

In eastern or southern Africa large baboon troops are difficult to count as they go about their daytime feeding, since they spread out over large areas. Yet, following a troop, I have often wondered if one might not work out a rough arithmetical key to the whole population simply by encountering the young of the same age and size. So tenaciously do they stay together that if one finds four, or six, or nine, the probability is strong that one has found them all. And while we may grant that in a given season there may be some variation in the number of newborn, or in their rate of survival, still nine of an age must indicate a very large troop. And they will all be there. Whatever may be the qualities of the fascination that unites them, play is its social function. And in the animal, education is play.

Victorians like Herbert Spencer tended to take a gloomy view of play, regarding it as what the irresponsible do with their surplus energy. Even the founder of modern sociology, Emile Durkheim, could find little better to say for it than that the world is not all play. Not till 1896, when a continental psychologist named Karl Groos suggested that play has survival value in that it prepares the young for mature tasks, did anyone seem to have anything theoretically good to say about it. Since then we have come far. In 1934 Carpenter recognized in his howler monkeys that juvenile play was second only to the mother as an instrument of normal development. His view was more than confirmed by Harlow's experiment with the rhesus orphans. Washburn and Rensch have each concluded that the slow growth of primate young would offer a selective disadvantage to species were it not for the longer opportunity to play and learn. We come today to the remarkable insight of J. M. Burghers that play is a game with the environment, linking the organism with the future, in which decision-making is exercised without too great a penalty for deciding wrong.

Preparation for adult demand goes deeply into play. Young patas monkeys may run thirty miles in a day; speed is their defensive resource when they mature. Kittens and lion cubs spend their days stalking and assaulting each other; excellence in the hunt will be important someday. Though forest chimpanzees nest at night high in trees, if there is danger they seek the ground; and the young play rarely in trees. Baboons, on the contrary, find safety in the trees and seldom move far from them; and young baboons are often in the boughs, chasing each other or playing follow-the-leader. I have suggested in another book that if human children play with guns, there is probably more to it than social corruption.

But there is more to play, also, than simple perfection of essential performances. William McDougall, an unfashionable name these days in sociology, saw play as preparation for adult competition, a comparably unfashionable conclusion. N. E. Collias found that even in play fighting, participants learn cooperation in keeping play play. Altmann invented the word "metacommunication" for the initial signal that all following screeches and yowls will be meant in fun. The adult human wink may be interpreted as metacommunication. Rules and regulations, and their acceptance, come about in play. The great Swiss child psychiatrist Jean Piaget, resisting the idea that children seek escape from rules, wrote: "Far from limiting himself to the rules laid down by parents and teachers, the child ties himself down to all sorts of rules in every sphere of activity, and especially that of play."

A sense of justice comes to us through play; we may think of young outrage when someone is accused of not playing fair. I was startled when first I read Suzanne Ripley's study of gray langurs in Ceylon, and found her description of monkey inhibition. Langur troops may be too small to provide enough young-of-an-age with a play group. An older juvenile may enter. But he will impose handicaps on himself to limit his strength and his roughness to the capacities of his juniors. Stuart Altmann, watching the descendants of Carpenter's transplanted rhesus monkeys on the West Indies island, found the same self-

limitation on the part of older juveniles in mixed play groups. Beyond that, he observed that play groups tend to be unstable unless games are "fair," unless approximately equal opportunity is presented for chasing and being chased, for tail-pulling and having your own tail pulled, for winning and losing.

As we penetrate more deeply the animal world of play, and

see it more clearly through young animal eyes, we come not only on the elemental importance of play, but on the elemental importance of the peer. Just as much as does the human child, the young animal lives in make-believe. Secret rules, secret signals, secret understandings are shared best by equals of age and experience. Out of the conspiracy the age group perfects the fantasy that is the play world of their waking hours. Emile Durkheim's "Life is not all play" would be echoed, I am sure, by the alpha rhesus signaling his status with upraised tail. The peer group, chasing one another around the social periphery, would not know what either Durkheim or the alpha was talking about.

Yet Jung's world of childhood fantasy must suffer its first collision with adult reality, whether human or rhesus or baboon, when the thrust of sex announces itself. Childhood is ended, adolescence begins. Few, however, are the primate species who, like man, have made a problem of it. And it may be that, just as with the declining influence of the human family we are turning to the primate solution of the age group, so some of us at least are turning to primate solutions to erase the problem of adolescent sexuality.

Chimpanzee or baboon, gorilla or vervet, the needy female takes on anyone who is interested. She is not undiscriminating. She tends to take the young ones first, subordinates of the establishment later, then perhaps on the fifth or sixth day of her estrus period to form an exclusive if brief consort relationship with the alpha. There is a suspicion that natural selection has acknowledged a subtle union of physiological and social relations for the benefit of the species. Her egg seems to descend and be fertilized, more often than not, in the period when she is with the alpha. If the conclusion is correct, then primate ingenuity is indeed a wonder.

The primate promiscuity of almost all species will be seen as an evolutionary advance only if we contrast it to what happens to subordinate males in so many lower mammal species. In Guhl's term, they are psychologically castrated. It is as if in primate species with increasingly complex societies and increasing individual diversity, a deal had been made. High-ranking males renounce sexual monopoly and sexual jealousy as a primate capacity, while in return maturing males, receiving the sexual franchise, agree to accept without too much fuss the exigencies of the social order. It is an accommodation between the mature and the maturing, a social contract. And while most obviously neither forgotten pro-simians nor evolution had it in mind when primate sexuality made its departure, still as a social innovation it has worked to preserve order in sexually rambunctious species. And it has preserved the individual, which psychological castration did not.

The young baboon, with adolescence, more and more becomes a part of the whole society. The play world slips away. Where peers once chased through the trees on the social periphery, as sub-adults they keep much the same range but now on the ground. They become the first line of defense, and if they still with their chasing spread far beyond the area of mothers and infants, then it is just as well. The more likely will it be that they will spot danger. Washburn and De Vore have written that a mixed group of impala, with their sense of smell, and baboons with their superb eyesight, is virtually impossible to surprise. And it will usually be these adolescents, out on the rim of danger, who provide the eyes.

On the same periphery one finds the four-year-old rhesus in process of integration into the establishment, but still sub-adult. John Kaufman, in another study of the descendants of Carpenter's transplanted monkeys, watched a peer group when their moment came to infiltrate the central area of hierarchy. They possessed no rank, of course, but only the omeganess of youth. Within any peer group throughout its growing up, however, distinctions will emerge, and leaders necessarily will have the advantage when the challenge to the establishment comes about. Nevertheless, when the seven challengers were resisted, they defended themselves in concert. The infiltration was in March. Into the summer they still acted together, but inequality was taking its toll. One was moving rapidly up establishment's ladder, but his fellows could not follow. Several found intermediate rank. But by September the peer group was gone, and the laggards had vanished into the omega throng.

They had fought for one another. Yet an inconspicuous justice remained. The one who rose so high was the princeling, the son of an alpha female whom I described in Chapter 4. And in two more years he was gone.

Human youth, turning from the disintegrating family to the communication and solace of its own generation, must reckon on factors both known and unknown in the world of the monkey. Human and animal must of necessity face one day integration into the larger society. Both, if the individual is to survive, must make compromises as a portion of the social contract to make survival possible. And boy or baboon, granted equal opportunity to display his worth, must be willing to settle for less. There is nothing new, in monkeys or men, about unequal performance.

But human youth, replacing the family as functionless, cannot substitute for it a functionless peer group. The young primate challenges, competes, plays fair, and learns. He prepares himself through a channel other than family for future contingency. If affluence has decreed that the human young will and must through innate longing accept the company of their peers as preferable to the company of their elders, they must— and I pray will—accept the obligation to learn which all animals have accepted.

The elders, of course, must create a human world worth striving for. Our youth has a right to ponder. But it commands no right to drop out.

3

If you are young, then a difficulty is that you do not know as much as you think you know. Out of such ignorance a youth movement may empower a Hitler, suffer enthusiastically the manipulations of a Mao, and ensure catastrophe for all. It may speak for anarchy in the voice of romanticism and bring to ruin the most advanced society vulnerable through interdependence. The rising power of the peer group, with its growing autonomy, introduces as a permanent feature of the contemporary landscape a monstrous truck, brakes gone, catapulting down a long mountain grade while smashing with no more discrimination than gravity itself all that comes in its way.

But if the maturing male does not know as much as he thinks he knows, the established male may not either. As ignorance is a property of the young, habituation is a property of the adult. We accept the way of the past as the way of the future. We follow obsolete road maps when the roads are long gone. We do not look below a surface to which we have become accustomed. We welcome lies we are used to, though we know them as lies; shrink from truths of strange dimension, though we must know them as truths. Change cannot be in the interest of the established; but then, neither can be social disaster, since the established have the most to lose.

Were our societies reasonably perfect mechanisms, then in their defense we might resort to any means to compel order and law. Among us may be those who indeed believe that we possess such credible perfection; and they may also conclude through some private mathematics that the world can find sufficient policemen to patrol all the side streets of youth. It is difficult to believe, however, that among us are many who have survived to maturity with perceptions so dim that we accept such mathematics; or who are not at least vaguely aware that something is wrong. Something, unconfined by national boundaries, undeterred by degrees of prosperity, unimpressed by political systems or ideologies, is wrong with human societies.

In the sciences we search for organizing principles, as the layman may search for a lantern to enlighten his dark misgivings. But so disparate are the illnesses of governments and men —of a Soviet Union, an Italy, a Britain, a France, an America, a Czechoslovakia, a China, a Japan—that no common prescription seems likely to be found. A coincidence, however, if it is broad enough, may sometimes bring on a wondering. And the investigator may stumble on such a puzzling coincidence: just as universal as the social malaise—as international, as apolitical, as divorced from levels of poverty and affluence—is the readiness of the young to revolt. Is there a clue here? Do the young know something that we do not?

It is a waste of time to ask the young themselves, for the cry will be "Everything!" and you will be showered with natural wisdom to embarrass a young Rousseau. The clichés of a century, all tried and found wanting, will descend on your head, and you will retreat with the conviction that if the young know something, then their demonstrations parade no symptoms. But the conviction is too easily bought. If the coincidence implies some common cause, some debilitating, universal germ, then why should the young more than we be in a position to describe it?

In late 1968 a clue came my way, and, as is characteristic of clues, it was of random origin. But it led to the discovery of another coincidence, one of chilling order, and while I am not in a position to assess its value, I believe it should be recorded. The accident was a letter from a reader. The clue came later.

Many a letter is intriguing. This was written by a Dutchman named Willem James, the manager of a large oil refinery in Rotterdam. He had introduced a new theory of management to his plant with consequences so successful that he found them difficult to explain. He wondered if the territorial principle might have something to do with it. I find it exhilarating to read, in our time, about something besides hardware that works. I was dubious, however, that the territorial imperative had been more than a contributing factor to his success. But I could be no conclusive judge. My normal paths of eavesdropping, wiretapping, and common detective work introduce me more frequently to herds of zebra and parties of waterbuck than they do to oil refineries. Territory, as things turned out, was to prove little more than a lucky link between animal species and human specifics.

A few weeks later, again by good luck, Willem James came to Rome for a company meeting. We negotiated a side meeting on the Via Veneto. He had brought a variety of xeroxed reports as well as his private testimony. And as I listened and looked, it seemed to me that I was in the presence of one of the more exotic stories of our times. But, studying his documents later and obtaining more material from the United States, I recognized that, far from an exotic tale, I was beginning to glimpse what I had never seen before: the hulking identity of the twentieth century's third man.

I shall compact the story. For far longer ago than our more paranoid youth would accept, industrial management has been worried. Why should the modern worker have a sense of being a bush without roots? A famous experiment—to become known as "the Hawthorne experiment"—was made among Western Electric workers at the Illinois plant. Various efforts had been made to increase the production of telephones—the company's technological destiny—without startling consequence. More money, more light, more space—nothing conspicuously affected production. Then a Harvard group under Elton Mayo was brought in. They isolated a few departments, introduced no dramatic changes, but emphasized that it was an experiment in which the workers were participating. Production boomed. Why?

That was in 1927. The experiment became famous. Management continued to chew its fingernails. And the soul-shrinking question to be asked is why did so few inspect the Hawthorne experiment to determine its psychic components? And of the few, why did almost none do anything about it? As I have previously suggested, capitalist and Marxist share the same *idée fixe* of the almighty dollar: that man works exclusively for reasons of economic determinism. The Hawthorne workers had been motivated by identity, not money—by being people different. They were incomprehensible.

Finally in 1960—and this is where James begins to enter the story—a professor of industrial relations at M.I.T. wrote a book called *The Human Side of Enterprise*. His name was Douglas McGregor, and he broached the theory that management throughout the world, regardless of political orientation, bases its strategies on a false assumption that man is lazy, shuns work, wants no responsibility, and reports for duty simply to avoid the punishment of material deprivation. He called it Theory X. One obeys out of fear, and so the management of men becomes possible.

You or I may recoil from such a generalization as Theory X. Can it be that all industrial enterprise of whatever shade of benevolence or ruthlessness, that all socialist or communist endeavors rest on such principles? What about Christmas parties, tea breaks, paid holidays on the Black Sea? Or is it all décor? And one recalls with a shock the environmentalist's dedication to security, the economist's presumptions of no want beyond material things. Overwhelmingly one recalls Pavlov's salivating dogs and Skinner's laboratory rats. We act as we do for no reason other than to secure reward and avoid punishment. An international, inter-ideological industrial superstructure has been shored up by the termite-ridden timbers of universally accepted behaviorist psychology. How can mere industrialists oppose "science says"?

Theory X presents the third coincidence: together with the human malaise and the revolt of the young, it passes beyond national borders, beyond political structures, beyond conditions of poverty or affluence. What if all three are related? And what if Theory X is correct?

McGregor said, No. He granted that it worked. The American economy could scarcely approach a yearly gross national product of a trillion dollars if it did not. But he maintained that it did not work well enough. He produced his Theory Y. And he turned to one of the heroic mavericks of American psychology, Abraham Maslow of Brandeis University, for a broader conception of human need.

At an earlier date Maslow had published *Motivation and Personality*, in which he presented his hierarchy of human need. Lowest and most fundamental came the physiological, then in ascending order came security, love or a sense of belonging, reputation or self-esteem, and finally, when others had been satisfied, the demand for self-actualization, for a man "to be actualized in what he is potentially." From it the late Dr. McGregor, to satisfy the whole man, put together his Theory Y of industrial management. Granted security, the worker was to be trusted, consulted, de-specialized, given every possible control over his own job, encouraged to learn others, allowed identity with the end product of his efforts. Theory Y implied the death of that most sacred of institutions, the assembly line.

Who would take a chance on a system like that? A man in California did. He was then vice-president of a small electronics firm producing high-precision instruments. His name is Arthur H. Kuriloff, and he faced too many defective instruments coming off the assembly line. The decision was made to adopt Theory Y. Management was reorganized to delegate authority down the ranks. Time clocks were abolished. The workers on the assembly line who had each learned and performed a small task were rearranged in teams of seven, each team responsible for producing a complete instrument, with distribution of tasks left to mutual agreement. At first there was confusion, helplessness, frustration. Then the more skilled taught the less, the helpless learned from the ingenious. Little over two years later Kuriloff could report in the professional journal *Personnel*, and later in his book *Reality in Management*, that man-hour productivity had increased by 30 percent, while reports by buyers of defective instruments had decreased by 70 percent. Absenteeism had dropped to one half the local average.

How much attention Kuriloff's experiment attracted, I do not know. Theory X remained in firm control of management the world around, and management remained worried, perplexed, and strictly doing business as usual. There were substantial questions. The California plant was small, its product complex. Large conclusions could not be drawn from an experiment so special. But it was then, in 1964, that Willem James made his leap in Rotterdam.

One can with difficulty imagine a less personalized installation than an oil refinery. His was fairly large, refining 80,000 barrels of crude oil a day. His dissatisfaction was entirely with efficiency and profits. He surveyed American refineries, studied Kuriloff's experience and McGregor's advice, concluded that a total reorganization—indeed, a philosophical reorganization—was necessary. While Theory Y was not entirely applicable to the operations of a refinery, he made his own adaptation. The hourly wage was replaced by a monthly salary. No checks were made on an employee reporting ill except to make sure that he was well when he returned. Since a refinery has no assembly line, the guiding principle became "management by objective." Wherever possible, groups were formed within departments and assigned an objective, but placed on their own as to the means of accomplishment. Advice was available, but management became more a consultant than a boss. The job was the group's.

"Job enrichment" was probably the most critical of the reforms, since as soon as a man had mastered a task he was encouraged to learn another. The more a man could do, the more money he made—but this did not seem to be the final reward. Adventure, challenge, the cause to dare was introduced into the life of the industrial worker. James told me in Rome, "The figures don't say it and maybe it sounds silly. We have a happy oil refinery. People come to work because they enjoy it." Yet the figures are astonishing. Manpower was reduced by 49 percent, productivity per man increased by 172 percent. If one goes back to a 1962 index of 100, then by 1967 production stood at 272. And labor turnover in these years decreased from 10 percent annually to 2 percent. It speaks much for James's happy refinery.

When I was a very young man I had the luck to meet the original Henry Ford. And in some manner that eludes description Willem James resembles him. But the one invented the assembly line and the other is destroying it. Ford was a rebel in American industry, since he paid his workers not what the labor market commanded, but enough to buy his product. Yet the assembly line denied innate needs. I recall, twenty years ago, a three-week experience in an aircraft plant in Texas recently removed, for strategic considerations, from Connecticut. There were seven thousand employees, and for reasons that need not concern us I had the freedom to ask whatever I pleased of the general manager or the floor-sweeper. It was a giant assembly line producing that most exquisite product, a carrier-based fighter plane. And I asked the personnel manager one day, "How could you have found in Texas seven thousand workers with such skills?" His answer demands long digestion: "Because there are only two hundred jobs that require more than six weeks training."

This is the assembly line, its limited demands, its limited rewards. It is likewise the assembly line of all organizations, mental as well as manual. You learn your job, and you do it. Theory Y denies all. Today McGregor's inspiration commands more and more attention. But it was Willem James's experiment in Rotterdam that proved that it works. Youth may regurgitate any notion that sheer profit can be the most revolutionary of motives. But we shall see.

Maslow himself once used a term of Ruth Benedict's which she had coined in a moment of rebellion against cultural relativism. "Synergy" describes the union of goals between the individual and his group. And synergy describes what has happened at Rotterdam. Theory Y has been proved a success. What will come of it? Perhaps nothing, despite all of management's conferences. Theory Y works. But labor unions dedicated to the cause of human mediocrity must and will oppose any change. Egalitarian intellectuals will join in the chorus. Yet, like the sight through a telescope of a planet and its satellites, the sight of a happy refinery in Rotterdam must give us pause to wonder. May not profits, like gravity, someday provide wonders?

I have not told this story to advocate an advanced approach to industrial competence. I do not know whether few, some, considerable, many or most of those enterprises consuming one third of the normal man's living hours are susceptible to the inducements of Theory Y. I do not even know whether few, some, considerable, many or most of the world's peoples have reached a degree of security permitting satisfaction of needs more profound. What I am permitted to suggest, within my own competence, is that a single experiment, broadly enough conceived and incisive enough in its outcome, can demolish a theory. And Theory X, though it be universally accepted, was damaged in California and demolished at Rotterdam.

From an evolutionary view I prefer my own triad of innate needs, applicable to all higher animals, to my friend Maslow's hierarchy of human needs. But they make in part the same statement: the pursuit of security bears reward only when you do not have it. And even possession of a slimmest portion may free us to seek impenetrable skies, imponderable answers, and a human dignity beyond definition. It is what youth in its ignorance knows, I believe, and we in our habituation do not.

Is this, then, the organizing principle of the triple coincidence of the international malaise, the international revolt of the young, and the international acceptance of the laboratory rat as the model for human aspiration? One encounters abruptly a fourth congruence: relative affluence. It is among those peoples and societies shaking off the bonds of poverty that we find not only the adult malaise but the shrillest cries from the young. We acknowledge it; we speak of the revolution of rising expectations. Our habituated minds, however, can interpret such expectations only in terms of material wants. But the uprising of the young has corresponded with perfect timing to the diminishing pressures for material security.

In our search for hypotheses of predictive value, I suggest that the concept of human organization motivated by material need has been sufficiently successful to destroy itself; and that if we do not enlarge our concepts of innate human need—our portrait of the human being himself—then our societies will eventually either lapse into apathy or explode into anarchy. An orthodox objection, whether by beleaguered management or bemused idealist, might accept the hypothesis but assert that the human being will not accept material sacrifice for immaterial gain. Yet the radical proof provided at Rotterdam denies material sacrifice. The refinery would not otherwise be in process of tripling its production today.

If we accept the hypothesis for its working value, then we may begin to comprehend many young displays: passionate concern for the individual and as passionate hostility for the organization; equivalent contempt for the labor union and the establishment of privilege; impatience with elder statesmen, older voices; equal disenchantment with political structures presumably individualist or presumably collective. We may even begin to comprehend why so often it is that the most intelligent among the young, the individual best equipped with perceptions to glimpse the universality of Theory X's degradation which maturity will force upon him, who accepts violence and the physical assault on social institutions as the only way: or turns to sexual abandon, to the kick of sensation, to drugs or other means of withdrawal, to decibels of entertainment too deafening to tolerate thought, and accepts the self-annihilation of the social drop-out as something that can at least be his.

If we follow my hypothesis that our needs are innate, and of animal origin, then we may likewise comprehend why the young need not know what they are doing to act as they do. Their drives rise out of the experience of species gone, and may be as obscure to the belligerents as they are to the besieged. Caught between the inadequacy of natural wisdom and the inadequacy of adult counsel, the most gifted spokesman for the young may dismay us with his ignorance, appall us with his heedlessness, divert us from responses short of apoplexy. And the irony is profound—indeed, it is a pathos—that the one counsel he accepts from us is false: the natural equality of men. Two centuries of lip service to a doctrine that no one ever truly believed has at last produced a generation who as true believers supervise their own castration. They condemn themselves to Theory X.

But do they believe it? My faith in evolutionary command commands me to doubt it. As youth, somewhere in his old-time bowels, rises in defense of the individual, that engine of primate supremacy, so I must suspect that our human brainwashing is not, as it has never been, a total triumph.

To take one last, sidewise glance at the inarticulate demands of the young, I shall turn in the final section of this chapter to the life of the baboon. Such an exercise is unflattering; it is also dangerous, for we are not baboons. Yet the ignorance of the young and the habituation of their elders may find some accommodation in the world of another species. Somewhere on the periphery of the baboon troop the young omega male barks sharply; the formidable alphas go to investigate. There is a cheetah, standing tall and thin on a high place, and he is looking at us. You and I will do well to look where our young are pointing. All comes more easily, one must admit, if you are a baboon and cherish no illusions, old or young, that you can survive without your troop. All comes more easily too, one must admit, for both alpha and omega if you possess no brain worth brainwashing.

4

The social contract of the chacma and olive baboons, in southern and eastern Africa, is the most severe in the world of the subhuman primate. Just as the citizen of a small, embattled nation will sacrifice a maximum of his civil rights to the order that he hopes will save him, so the eternally embattled baboon sacrifices a maximum of animal freedom to that social order which permits his survival. Yet he sacrifices little individuality.

The baboon is the largest of monkeys, and differs from the ape most conspicuously in that he boasts a tail. His largeness, however, is what invites the first of his vulnerabilities. All monkeys are edible, but most are too small for predators other than snakes and jackals and eagles to bother with. They are tidbits, and uncomfortably agile tidbits too. But the baboon's size, while offering unimportant temptation to social killers like the lion who must have a large animal to satisfy many hungers, comes very near perfection in cheetah or leopard estimate. Wherever men have killed off leopards, baboons have enjoyed a population explosion.

The second of baboon vulnerabilities is his life on the ground. Primate students are coming more and more unanimously to the conclusion that terrestrial life has commanded increased aggressiveness in the primate individual and tighter organization in the primate society. Life on the ground (and this is of untold significance concerning the evolution of the human primate) offers hazards beyond measure as compared with a life in the trees. Such a survival disadvantage has a monkey on the ground that the observer must wonder why anybody ever tried it. But baboons did, as did the human ancestor and many another. We shall wonder why later; now we shall simply observe the consequences. The edibly attractive baboon, whenever it was that he could no longer resist terrestrial temptation, became a famous target.

For the third of baboon vulnerabilities the baboon has no one to blame but himself. He cannot keep his hands off other people's goods. In the earlier millennia of his evolution, edibility was enough of a problem. Then came man with his weapons. In many a primitive African people the young proved their manhood by killing a dangerous animal such as one of the great cats. The baboon should have greeted man as a savior; he greeted him instead as a sucker. Baboon temptation could not resist the patches of maize grown by African farmers, as later it could not resist the fruit of the white man's orchards. The appetites of diminishing leopards and cheetahs were more than replaced by the angers of multiplying men.

The baboon is at war with the world, and has been so for time without known beginning. Beset on all fronts through an eternity to make the siege of Leningrad seem the flicker of an eyelid, baboon defense has been the order and concert of numbers. The authority and responsibility of the few, the loyalty and obedience of the many, have through group selection perfected a social contract that works. There are today, against all odds, more baboons in eastern and southern Africa than there are people.

If the baboon contract is severe, then we are permitted to understand. If order weighs heavily over disorder, then the balance could not be otherwise. The rule of the alpha oligarchy may be unchallengeable; punishment for the delinquent heavy. Ruthlessness of leadership may be such that the sick or disabled must struggle frantically to keep up with the moving troop, since to be left behind is to become a dinner for the nearest predator. This is baboon life. Yet in all measures of personality—alertness, diversity, aggressiveness, ingenuity—baboon individuality remains. How have they done it?

Again let me consult my triad of innate needs, common, I have suggested, to men and all higher animals. There is identity, as opposed to anonymity; there is stimulation, as opposed to boredom; there is security, as opposed to anxiety. That the baboon troop has effected adequate security for the individual finds its proof in the survival of species. Yet the security is that of the group, and the security of the individual is that of his identification with the group. In the face of hazard, all survival rests on individual resource: the alertness of the young, the willingness of the alpha to accept risk, the fidelity of the mother to her infant. Washburn and De Vore reported that a baboon without a troop is a dead baboon. And it is as if, buried in the baboon subconscious, the truth, like one's shadow, can lie never far away. Security for the group becomes stimulation for the individual.

Interlocked in the baboon contract are identity, stimulation, and security. One cannot be insecure so long as identity remains solidly with the troop; and one can scarcely be bored if responsibility in the face of hazard falls on one as it falls on all. The contract implies, most obviously (though no more obviously than in man), recognition of what happens to a baboon without a troop.

In such fashion are innate needs satisfied by response to hazard. But all life is not hazard. And baboons retain the integrity of the group and of the individual in the less threatening avenues of the game reserve. Here man ceases to be a threat. Daily life consists of feeding, playing, grooming one another, dozing through the siesta hour, an occasional bout of sexual entertainment, an occasional quarrel broken up by the alphas, more feeding, and the dusky return to a cliff, if you have one, or, if not, to your sleeping trees. There is little apparent excitement.

Even in a protected area, however, there remain those enemies of older claim on baboon antipathy than men. Alertness cannot be dispensed with. One night in a camp at Lake Manyara, a small gem of a reserve lying in the bottom of Tanzania's Rift Valley, we were awakened by baboon ruckus. The valley wall beside us was 2,000 feet high, the same wall that the elephants climbed, and a troop had a customary sleeping place in rocky cliffs several hundred feet up the cliff. The clamor of barking was of an order to be inspired by none but the leopard. We detached a portable spotlight from a truck, aimed it up at the valley wall. We could not see the bodies of the prowlers, only pairs of green eyes reflected in our light. It was a night of remarkable leopard sociability; two were hunting together. A pair of eyes would vanish for a while, probably behind a bush, then reappear. Baboons, unlike chimpanzees, do not vanish into darkness, silence, and personal safety. The racket became a deep uproar, the uproar a shrill cannonade. After a while we lost the green eyes. After a much longer while the clamor subsided to infrequent barks, then to the silence of the troop again sleeping in its cliff. Whether the leopards had retreated in discouragement or had taken a baboon or two and gone off to feed, we could not know. The baboons could not know, either. We too went back to sleep. Such is night in Africa.

And so must always be the life of the baboon, until poachers have taken the last leopard to be fashioned into an expensive coat. How will he manage at the heart of a rich, protected area in a world without want or danger? Would his innate needs yet be satisfied? I believe they would. And the secret of his success would lie in that undistinguished, unwashable brain.

The baboon will never persuade himself that aggressiveness is a product of frustration. The young will never blame their failures on lack of parental love in infancy, as adults will be unlikely to forbid the young their tendency to play in trees. Should the proposition that competition is somehow wrong come baboon way, small brains would be dumbfounded; should some mutant baboon idealist insist upon it, he would be greeted with the lifted eyebrows not of human surprise but of monkey threat. That baboons in their original state had been happy and good would create little but embarrassment for ancestors so dull; that society has brought baboon downfall would be greeted with a chorus of protest, for without society how would anyone learn even what to eat? The mutant idealist, despite all threatening eyelids, persists: there must be a declaration of the natural equality of baboons. He will receive the ancient roar reserved for the extinct leopard as a natural enemy of their kind. For who wants it? The young baboon? He demands the freedom to assert if he can his natural superiority. The mothers? They are busy enough with their infants, further responsibility is outrageous; and bleak indeed would be the sexual future without that breath-taking crack at an overwhelming alpha. The omegas turn to such displacement activities as digging roots; life is far more comfortable being a non-equal. The middle-class males are confused; what would they do for excitement without status arguments? The alphas sigh, retreat beyond a few bushes, and start up a crap game.

Baboons—out of intellectual limitation, perhaps, and most certainly out of an incapacity for self-delusion—are unlikely to surrender a society that fulfills their innate needs. The fundamental demand for security will be answered so long as the troop acts as one. But security cannot be transmuted to boredom so long as competition, free but subject to equal rules and regulations, invests the chasing peer group, the prospects of the maturing male, the jockeying for position between Number Eight and Number Nine, the sexual ambitions of the female in estrus working her way up the social order. Stimulation—the excitement of victory within the rules of the game—is written large in the rewards of baboon society as in the non-existent society of undeluded men.

It is identity—what Maslow would call self-actualization—that presents the most incisive challenge to the habituated human mind. Who am I? Where am I? Have I actualized myself, singularized myself as my genetic potential proposes? Why a Douglas fir will grow to its maximum height, why a chicken will struggle for her highest possible rank in the pecking order, why a man will sacrifice fortune for fame, why the aspiring leader of a juvenile gang will neglect all safety to retain his status, why a criminal will glory in a score none other would have dared to attempt—all are avenues to identity in one's own eyes and the eyes of others.

But identity does not demand alphaness. Neither you the reader, nor I the author, yearn desperately, I suspect, to be President of the United States. Like many another being of lower rank, we may behold with wonder, with relief, and perhaps with amusement the insane efforts of some fellow citizen to get the job. If we shelter a personal hope, it is that the selected alpha will generate a following response that will affect us. For herein lies another powerful ingredient of identity, another means of answering the undying question, Who am I? To be so privileged as to point to a man with pride and say "That is my husband" or "There goes my boss," or to include a group, perhaps with switchblades shining, and say "This is my gang."

And still, identity does not stop there. For rank order in itself provides fulsome means of identification, and perhaps this is one reason why rank has been such an evolutionary success. One need not be alpha to know who one is. Number Six will do. Every rung on the social ladder carries its identification, and is a step higher than somebody else. There is always the omega, of course, but he seldom cares, and, like the village idiot or the town drunk, acquires his own special character and identity. If the omega cares, then there is probably something wrong with the system.

Many another means of acquiring identity is offered to the world's beings: There is association with a place uniquely one's own, like the starling's perching place, the Peruvian viscacha's sunning place, the topi's defecating place, the hartebeest's sentinel place on a termite mound. There is the defended territory of the Australian bower-bird, of the northern roebuck, of the Asian gibbon, of the African vervet-monkey band, offering not only identity of place but stimulation of defense and frequently security of food supply too. Place enters profoundly into identification, whether it be that of the drunken Southerner weeping in his cups at a strain from "Dixie," the dog returning home despite all his master's efforts to give him away, the Pacific salmon returning after years at sea to that little stretch of brook where he was hatched, or the seemingly superfluous identification of place by a painter called Leonardo da Vinci. Strange hobbies, strange religions, strange love affairs, strange dress—any of them may reinforce our satisfactions of uniqueness. But as we must seek identity as we seek the sun, so we must dread anonymity as we dread the dark.

The young baboon matures with the assurance that in his natural society he will find a place unique and his own. The human youth has no such assurance. A tiny aircraft facing the dense, implacable wall of a cold front may turn back; but life is irreversible. Like some monstrous whale devouring plankton by the acre, so the organization of modern life devours the individual. Specialization will reduce him to a needle lost in the twentieth century's organizational haystack. Classification will place him with all beans of equal size. A workingman, he will be denied the right of excellence by his union. An organization man, he will be required to deposit his most secret riches in the company safe. His life is ordained. From an anonymous house in an anonymous suburb he will take his anonymous seat in an anonymous train to reach an anonymous office where he will perform the tasks he performed yesterday, eat the lunch he ate yesterday, suppress the resentments he suppressed yesterday, return to his anonymous seat in an anonymous train to an anonymous suburb and an anonymous home where he will have the martinis he had yesterday, renew the quarrels he had yesterday, watch lights and shadows on a television screen indistinguishable from the lights and shadows he watched the night before. And he will go to bed. A morning will come when he will not rise. His life, while it lasted, has of course been secure. And to a degree, one must confess, an unnatural society will have achieved among its members a measure of unnatural equality.

Human youth recognizes that a few achieve identity. But it is a shrinking few, as organizations devour each other, while youth grows in numbers. And so there are those among the young—today some, tomorrow more—who suggest that if something does not give, then they will tear the place down as a house not worth living in. There is nothing unusual, in the quest for identity, to find those who will contemptuously reject security's last offer.

6. Death by Stress

In 1932 the director of the New York Aquarium, C. M. Breder, Jr., working with a colleague named C. W. Coates, performed an experiment with those small fish known as guppies that shook no worlds. Their conclusions were published in a little-read scientific journal, *Copeia*, and few people today are aware of their work. Yet the fifty-one guppies who participated in Breder's adventure should one day be memorialized by some watery monument, for they not only discredited a Pope but threatened with ruin a scientific doctrine as unquestioned as any in our time.

There are few of us unfamiliar with the tiny fish so common in our children's aquaria. Guppies multiply lavishly, and are born in a ratio of two females for every male. Breder initiated his experiment by arranging two tanks of equal size, each with an abundant food supply and aeration ample to tolerate a host of fish. Then in one tank he placed fifty guppies with an unnatural distribution of approximately one-third males, one-third females, and the remainder juveniles. In the other tank he placed a single gravid female—one heavy with eggs already fertilized. What he expected to happen, I do not know. What happened defied prediction then, as it defies explanation today.

A remarkable character of the pregnant female guppy is that a single fertilization may give as many as three broods, born every twenty-eight days. The lone gravid female cooperated nobly with the experiment, producing broods as high as twenty-five. Yet at the end of six weeks there remained nine fish in her tank. She had eaten the surplus young. In the meantime the tank with an original population of fifty had witnessed a rapid and immediate die-off. Cannibalism of newborn was so rapid that it was seldom witnessed. The fish surviving at the end of six weeks had all belonged to the original population. Here too there were nine. In each tank there were three males and six females, the ordained proportion among guppies.

An ironic turn in the history of science took place when both Charles Darwin and Alfred Russel Wallace found their inspiration for natural selection in Malthusian doctrine, a thesis which sooner or later must be accepted as in large part false. Thomas Malthus was an English economist who in 1798 published his *Essay on the Principle of Population,* demonstrating that while human populations increase at geometric pace, food supply can increase only by addition. The multiplying population must therefore at some point overtake the supply of food. And at that point the population will reach its limit.

Darwin and Wallace saw in the Malthusian doctrine a natural law which must apply to all species, and so they deduced that through competition for a limited resource, food, selection must take place between fit and unfit. The Malthusian logic seemed inarguable, and, as we know, upholds the pessimism with which many view our contemporary population explosion. And undoubtedly supply of food places a theoretical limit on animal numbers, just as there must be cases in which deficiencies of quantity or quality of food contribute to a limiting effect. Yet the new biology provides no proposition more demonstrable than that of the self-regulation of animal numbers. Rare is the population that has ever expanded until it reached the limits of food supply. Rare are the individuals who directly compete for food. An infinite variety of self-regulatory mechanisms, physiological and behavioral, provide that animal numbers—except in the case of climatic catastrophe—will never challenge the carrying capacity of an environment. Birth control is the law of the species

When Paul VI in 1968 shook the world both Catholic and non-Catholic with his condemnation of contraception, he made an error of fatal dimension. Had he and his advisors confined their condemnation to one of conflict with church doctrine, then a student of evolution could have nothing to say. But contraception was also condemned as a violation of natural law. And, whatever private interpretation the Vatican may give to the phrase "natural law," its invocation permits debate by the observer of natural arrangements. The conclusion which I shall present, drawn from the evidences of the new biology, is that contraception is the cultural implement enforcing a natural law. Papal condemnation, not contraception, stands as violator of natural designs.

In terms of the social contract, we may say that just as society must furnish the young with freedom to develop their genetic potential, so the breeding adult must not provide society with more young than the group can handle. The propositions are poised in equity, and the neglect of one must result in the nullification of the other.

Fifty-one guppies, controlling their numbers through a blend of cannibalism and infanticide which we must assume seems quite normal to guppies, can scarcely be regarded as furnishing a sufficient case for the toppling of a Malthus, the indictment of a Pope, or the elaboration of a social contract. But other evidence exists. And as we review the efforts of scores of scientists in recent decades, and as we identify ourselves with their successes small or spectacular, their failures comical or maddening, we shall gain, I believe, a proper mass of evidence. And we shall enjoy, too, participation in one of science's better detective stories.

2

Immediately after the First World War a burst of originality possessed us. Perhaps trench warfare, with its massive insult of collective death, brought to the human spirit a demand for individual daring. The literary consequences are famous, for these were the years of the uninhibited Left Bank, of Joyce, of Hemingway, of Fitzgerald. It was the time when unknown playwrights were popping up all over the Times Square area to create in the American theater its one golden age. Jazz overleaped all language barriers, and the American popular song became the musical currency of far markets. And in the sciences originality too was displaying its wares.

Out in Chicago the psychologist John B. Watson was founding behaviorism on the principles of Pavlov's conditioned reflex. His effort would wind up with today's reinforcement theory, and while I may deplore Watson's contribution to scientific confusion, I cannot dismiss his daring. Less conspicuously Eliot Howard, in England, was presenting us with the concept of territory, while in Norway Schjelderup-Ebbe's observation of barnyard pecking orders complemented Howard's to give us the foundations for a future ethology. Meanwhile the University of London's Carveth Read was publishing his *Origin of Man* with its prediction that the human ancestor, when found, would resemble in his life style more the way of the carnivorous wolf than the way of that inoffensive vegetarian, the forest ape. And at the same time, in the faraway Transvaal, Raymond A. Dart was discovering Read's predatory fellow in the person of *Australopithecus africanus*.

Most of these bursts of originality almost half a century ago disappeared for decades, leaving small trace. None went into a hiding more perfect than Sir Alexander Carr-Saunders' *The Population Problem*, an overwhelming attack on the Malthus doctrine. That was in 1922, and it failed to start even a back-room argument. But, fortunately, just two years later a paper appeared called "Periodic Fluctuations in the Numbers of Animals," by the pioneer ecologist C. S. Elton. It threatened directly no sacred assumptions, but it initiated a line of research that has progressed ever since. While Elton's study concerned largely the snowshoe hare, it was inspired by the lemming, an animal whose inexplicably suicidal tendencies qualified it as scientific box office. The rumpus was on.

The morbid activities of the lemming had been tantalizing human imagination for centuries. Travelers brought back grisly accounts of dead lemmings floating along the Norwegian coast. The earliest report to come to my attention was made by a Jacob Zieler in 1532. The first true study was made by a naturalist named I. G. Gmelin while observing animal populations in the Siberian tundra in 1760. He discovered the regularity of fluctuation in the numbers of both the lemming and the Arctic fox that preys on it. When the time of vanishing arrived, lemmings simply disappeared into the Siberian immensity. But in the confines of Scandinavia they did not always disappear. A ship off Trondheim in 1868 reported a shoal of lemming corpses so enormous that the ship took fifteen minutes to pass it.

As has frequently happened in the contemporary studies of animal behavior, the activity of a single species has led scientists on to broader and broader research, with conclusions of deepening significance. Carpenter's study of the howling monkey introduced us to the intricacy of primate society in a state of nature. Howard's warblers have given us today the territorial principle. The behavior of a few barnyard hens has led us on to a recognition of rank order as an organizing structure in most animal societies. Konrad Lorenz' imprinted ducks have brought into question all those learning theories dear to behaviorist psychology. G. T. V. Matthews' preoccupation with homing pigeons, initiating a wave of investigation into animal navigation, has left us struggling with the problem of just how much we know about anything. If traditionalists have resisted the chain reactions of ethology, it has been with true intuition that their tidy explanations are in danger.

So the lemming today joins the ranks of the classic animals. A tiny beast weighing only a few ounces, it is a member of that largest of all mammalian families, the *Muridae*, including mice and rats, hamsters, gerbils, voles. And if the lemming's importance to man is minimal, then this is only because while existing in probable trillions in the sub-Arctic wastes, it exists where men are few. Yet the chain of investigations initiated by the lemming's bizarre behavior is taking its arrow's course toward a target of human affliction.

Why do the numbers of the lemming build up every few years, then crash? Sweden witnessed in 1963 one of the greatest "lemming years" in decades. Lapps in the far north first reported the disappearance of lemmings in the month of August. Originating in the mountains, they vanished, moving south.

They moved mostly at night, and observers at a crossroads counted forty-four pass per minute. They moved as individuals, not as groups. Although food was abundant, if one died he was immediately eaten by others, the skull being opened neatly and the brain being eaten first. Of several hundred taken and examined, all proved to be the young of the year, and although sexually mature, not a female was pregnant. The migration was a youth movement. At any water obstacle, like a lake, they massed on the beaches in such number that an observer could not move without squashing them. Fair enough swimmers, they were not good enough. From a dead-end peninsula on Lake Storsjon so many obeyed the unknown impulse compelling them that the shores of the lake, the following year, were carpeted with lemming bones.

Darwin, without the benefit of future observation, guessed that recurrent epidemics caused the periodic die-off. Through the years, however, most observers agreed that for reasons probably of food shortage, lemmings stage regular migrations out of an area. In Siberia's endless wastes they go elsewhere, whereas among Scandinavia's lakes and fjords and seas migration becomes suicide. The answer seemed good enough until a skeptic in 1921 demonstrated that, whatever the place or the environment, lemmings vanish everywhere simultaneously, appearing nowhere else. Then in 1924 Elton said: sunspots.

The answer might seem a touch cosmic. But what Elton had done was to correlate Scandinavia's peak lemming years with those years in Canada when snowshoe hares appeared in greatest abundance and staged similar population crashes. The years were about the same. Then, with a mighty stroke of ingenuity, he went back through the yellowing records of the Hudson Bay Company. Since the eighteenth century the company had kept season-to-season records of furs purchased. Lynx purchases, following the same peaks and valleys, had varied from 3,000 to 80,000.

Even more exclusively than the fox preys on lemmings, the Canadian lynx preys on the snowshoe hare. The great naturalist Ernest Thompson Seton once wrote that the lynx "lives on Rabbits, thinks Rabbits, tastes like Rabbits, increases with them, and on their failure dies of starvation in the unrabbited woods." Seton in 1886 estimated the number of hares as 5,000 to the square mile, yet after a crash there might be only one. Elton found that Seton had been right about the unrabbited woods, for pelts of the snowshoe hare in the Hudson Bay Company's records swung between 10,000 and 100,000, in accordance with the cycle of the lynx. And that cycle varied between nine and eleven years.

Now Elton had his second inspiration. The lynx-hare cycle corresponded with the cycle of sunspot activity. His investigations indicated that the lower the activity of solar flares, the colder and dryer the climate on earth. (Recent studies of British Columbia forests confirm that minimum timber growth corresponds with minimum sunspot activity.) Elton reasoned that in the marginal conditions of the sub-Arctic, declining sunspot activity would have its most drastic effect. If in good years fecund animals like the hare or lemming built up to huge numbers, then deteriorating climate and food supply would bring on the consequent decimation.

In the most funereal mood one must record the death of a hypothesis so brilliant. Some years later a Canadian scientist, D. A. MacLulich, a man one must assume quite devoid of compassion for fellow scientists, through most studious checking found that in the 174 years between 1751 and 1925 there had been eighteen lynx-hare cycles, and only fifteen sunspot cycles. The final crash was that of C. S. Elton.

And yet MacLulich's victim was indeed not the ecologist whose originality had been of such order as to make even being wrong a triumph. The victim was the principle that it is food supply that limits population, and the corpse being secretly carried through science's back streets was that of Thomas Robert Malthus. A later authority, Dennis Chitty, could write just a few years ago: "No animal population continues to increase indefinitely and the problem is to find out what prevents it."

The food theory died hard, if it can be said to be dead even now. Kalela, a Finnish ecologist, conducting a three-year study of the red-backed vole in Lapland, recorded a devastating crash in a season of abundance. And yet in the same year a University of Helsingfors biologist could publish summaries of cycles of the Arctic fox and ptarmigan in Greenland, the partridge in England, the red grouse in Scotland, the goshawk and willow grouse in Norway, and the waxwing and red fox in Finland, to prove that all was climate. Chitty himself made an elegant study of the vole in Wales. The decimation of one population coincided with bad weather in 1938; but an adjoining population did not suffer until the following year, when the weather improved. A decade later, in the succeeding crash, neither weather nor food shortage contributed influence.

An experiment in 1965 with the California vole recalls Breder and his guppies. Two areas, each of an acre or so, were isolated. On the first was a population of voles increasing naturally at a rate of about 3 percent a month. On the other was a slim population of five males and eight females, and this became the experimental area. High-quality food was periodically scattered on it. Fertilizer was applied to enhance the natural growth. The population boomed to a springtime peak of forty-seven males and fifty-three females. But by August it had returned to precisely the original number—five males and eight females.

Biology's nineteenth-century certainties were being replaced by the twentieth century's open questions. If it was not food supply that limited animal numbers, then what was it? The famous Iowa ecologist Paul Errington became absorbed by the problem. His long career, first as a hunter and trapper, later as research zoologist and teacher, denied him allegiance to orthodoxies. Errington's was the kind of mind that could distill long observation to primary-school simplicity: Muskrats, grouse, ring-tail pheasants, and snowshoe hares all reach a maximum population in years ending with 1 or 2, a minimum in years ending with 6 or 7. Mathematics, not environment, determines population. He recalled further Hudson Bay Company figures. In the mid-1930's a huge engineering project in Saskatchewan affecting three quarters of a million acres provided as a side effect more and better habitats for muskrats. The fur catch increased in consequence. But though the base was now higher, the fluctuations proceeded just as before.

An argument frequently advanced has been that the increasing number of predators—as in the lynx and hare or in the fox and lemming—at last decimates the prey population. Errington dismissed it. The traditional enemy of the muskrat is the mink, as savage a little killer as the world affords. When the muskrat population is at the peak of its vitality, the most formidable attention on the part of the mink results merely in mother muskrats having more young. "Diminishing of population tensions affords one of the best stimuli for reproduction." Anything, in other words, that reduces the density of a healthy population, whether predation or drought or epizootics (that admirable word for epidemics so standard in zoology's vocabulary), will increase the birthrate in compensation. But mathematics must favor the mother. Let drought fall in a year ending in 7, catastrophe may ensue.

Errington concluded that much in the resilience of natural populations lies beyond present knowledge, and that unknown factors may depress certain life processes. The Bergsonian conclusion carried little appeal to a scientific community which tends to confuse the insoluble with the insulting. But his concern with population density and its relation to reproduction, recorded in a paper published in 1951, contributed to a course that investigation has followed ever since. Muskrat litters vary in number, as likewise do the number of litters in a given season. In a time of low population density the mother muskrat may have as many as twelve to sixteen young in a year; in a period of topheavy numbers as few as two or three.

How self-regulation is accomplished remained as great a mystery as ever, but it related directly to density. The new clue exploded innumerable investigations. Somewhere in the population haystack was a needle, and in the scientific night one spied a host of detectives, each with flashlight, magnifying glass, and Sherlock Holmes cap, trampling through the hay with his shoes off. But to qualify for such a role of scientific private eye, you needed a combination of resources seldom known in the sciences. You needed the experience of ethology, with its emphasis on the relation of animal to animal. You needed the viewpoint of the ecologist, with his principal attention on the relation of the animal to its physical environment. And you needed command of physiology, with its studies of organic processes. If you were lucky, you knew about evolution too, and the relation of the past to the present. Out of all such exhausting demands emerged a new discipline called population dynamics. Few could qualify. Yet the hay suffered little from neglect.

In the following section of this chapter I shall explore the self-regulating devices which in many species prevent a build-up to such numbers that only a population crash can provide natural limitation. The species we have been considering—in a sense, the freaks—lack such mechanisms and so are subjected to cyclical control. And while there must remain a temptation to equate man with the lemming and the snowshoe hare, we had best defer our consideration of men until we know more about animals. And we had best defer too those questions facing the new devotees of population dynamics in the 1950's: How, when density reaches a certain point, does a form of birth control take place so that fewer young are born, or are even conceived? And why—a more difficult question—do the elders drop dead?

The mystery of death and life, brought to attention by the insignificant lemming, has scarcely been resolved by a Vatican decision.

3

John Calhoun, a director of our National Institute of Mental Health, is, like Maslow, a maverick's maverick in the field of psychology. Physically slight, temperamentally elusive in the sense that elves are hard to get hold of, Calhoun is blessed with the capacity of slipping through the formidable fences of American psychology to escape without attracting undue notice. His first escape of which I have record was in 1952.

For two years the psychologist had been observing a population of wild Norwegian rats which he established in a large pen at Towson, Maryland. So large was the pen—almost a quarter of an acre—that under laboratory conditions of small separate cages it could have accommodated 5,000 healthy rats. At Towson, however, the undomesticated rats lived under approximately normal conditions. Calhoun established his population with five pregnant females. In two years the theoretical number of possible descendants would be 50,000. Yet the population of the pen never exceeded 200, and gave every indication that, no matter how long the experiment might be extended, this would remain the approximate number.

As in Breder's experiment with the guppies, final numbers bore relation not to food supply but, in some fashion or other, to space. Mother guppies controlled numbers by the indelicate process of devouring surplus young. Calhoun's young suffered to a degree through adult fighting and neglect of maternal care. But as the population stabilized, it became evident that the critical factor limiting numbers was territorial behavior.

Norway rats form stable societies when no more than a dozen adults share a territory and jointly defend it. Within this little world adults form a hierarchy led by an alpha male. The amity-enmity complex which I described in *The Territorial Imperative* turns hostility outward and preserves peace within the group. Calhoun wrote that the territories, and the buffer zones between, "seemed essential to the maintenance of group integrity." But they likewise divided up the available space into homesteads for groups of limited number. So population control was achieved.

It was the Irish ornithologist C. B. Moffat who first glimpsed in territorial behavior a means of controlling animal numbers. That was in 1903, long before Eliot Howard introduced the whole territorial principle to biology. A portion of the earth's surface exclusively your own brings you as the proprietor many a material benefit. You are defended, since you know it better than do your enemies. Possession in some strange way enhances your energies. Through a process of animal justice, might no longer makes right and on your home grounds you are capable of resisting intruders stronger than are you. Territory may ensure a food supply for you and yours. If you share the property with your wife, then it will ensure also that you do not leave each other when the children need you both. These are benefits accruing to the proprietor, but there are two powerful benefits accruing to the population and the species. By the physical separation of individuals or groups, dangerous aggressive forces are reduced to shouted insults over common boundaries. And the distribution of available space among breeding couples or groups means that the number of offspring will be kept well below the carrying capacity of the environment.

Although with our present knowledge of the territorial principle Moffat's insight might seem self-evident, still only in recent years has the relation of territory to population control become accepted. Even in 1956 Cambridge's eminent ethologist Robert Hinde rejected the proposition as unproved. His rejection inspired at least two studies designed specifically to demonstrate limitation of births through territorial behavior.

Hinde's objection had real grounds: While the necessary possession of a territory quite obviously distributes breeding pairs throughout an environment, it does not follow that territory limits the numbers of such pairs. Space is seldom that scarce. The English robin, for example, must have its breeding territory, but there remain in England vast areas of suitable habitat unfrequented by robins. And so to meet the objection Adam Watson, a colleague of V. C. Wynne-Edwards at the University of Aberdeen, set up an experiment with red grouse on the Scottish moors. And David Carrick, of Australia's Division of Wildlife Research, began his observations of the Australian magpie in fields near Canberra.

The problem was to demonstrate that the territorial necessity actually eliminates healthy adult birds from the breeding population. Red-grouse males establish territories in the autumn, holding them till the following summer. Space for breeding in the moorland is truly unlimited, yet the competition for territories takes place only in restricted areas. Watson began by

selecting a study area and marking all birds in the vicinity. He then cleared 119 territories by capturing or shooting the proprietors. Within a week 111 were filled by new males, only a dozen of whom were of unknown origin. All the rest had come from the marked population of the vicinity which constituted a nonbreeding reserve. All bred successfully the following spring. Watson had demonstrated that at least in the red grouse it is the number of breeding territories that in fact limits the breeding population.

The Australian magpie is related to the crow, but it is a creature quite distinct in both appearance and behavior. Like the crow, it is a large bird but with vivid white markings on the rump, at the nape of the neck, and in the flash of outspread wings. Some, again like the crow, gather in large non-territorial flocks; but others form social bands surprisingly like Calhoun's Norwegian rats. Up to ten adults defend as a team a property of five to twenty acres. And these are the only birds that successfully breed.

Carrick's study area was about five square miles of savanna, broken here and there by eucalyptus clumps, supporting thousands of magpies. Both food supply and nesting sites were unlimited. A few non-territorial groups nested in trees, but never succeeded in raising young. The crow-like flocks in the fields never tried. Successful breeding was confined to that 20 percent of the total population within the territorial bands.

What, precisely, prevented normal reproduction in any but the propertied bands? Carrick made a surprising discovery. While all males in the total population produced motile sperm, only the hens in the territorial groups ovulated normally.

The physiological link between normal sexuality and territorial possession is demonstrable in many species. The most prevalent of the many arrangements was first conclusively demonstrated by Eliot Howard's observations of countless species of finches and buntings, warblers, lapwings, woodpeckers in which the female is sexually unresponsive to an unpropertied male. The Marxist mind may be properly revolted by such conduct, and gloomily conclude that the study of animal behavior can come to no good. But the way of the lapwing persists. Furthermore, later research demonstrated the converse of Howard's proposition: the unpropertied or dominated male tends to be psychologically castrated. Thus not only does female sexuality guarantee that breeding will be accomplished within a select circle, but sexual inhibition provides that the disenfranchised male will break up no homes, indulge in no rape.

We still have no certain answer concerning the physiological linking of territorial behavior and the sexual impulse. But the field research proceeding today leaves no doubt about its reality. In 1966, when I first investigated the territorial imperative, I could find only one species of African antelope to which I could responsibly ascribe the pattern earlier observed in birds. This was the Uganda kob, a species in which males occupy an arena of territorial competition to which females are attracted for copulation. Females will accept no other than the successful, and the masses of surplus males amuse themselves in their bachelor herds. In the few short years since I published my review, territorial systems of breeding have been described in the wildebeest and waterbuck, in the Grant's and Thomson's gazelle and the comparable southern springbok, in the hartebeest and topi and puku, and in the smallest of them all, the oribi and dik-dik and steenbok. Systems vary, from the modified arena competition of the wildebeest and puku, to the bird-like family territories of the steenbok. But, in all, the main propositions hold true: The female will be attracted only by a territorial male; the male who has failed in the territorial competition will retire into the careless existence of males in groups.

As we are only now discovering the prevalence of territorial sexuality in the large family of antelope species, we stand only now on the threshold of discovering the undoubted physiological explanations. But in our expanding efforts we come on some strange behavioral mechanisms of population control.

The wildebeest, one of the most common and certainly the most grotesque of African antelopes, possesses no social organization worthy of analysis excepting the incidence of territorial bulls who maintain a monopoly on copulation. Beyond that, wildebeest, like schools of fish, congregate in immense herds, offer the lion his favorite dish, and migrate with the rains in the hundreds of thousands. And population control takes place in an adjustment of unlikely instincts. The mother drops her calf without more emotional engagement than might take place with a bowel movement. But the calf has a following instinct regarding its mother. Miraculously, the newly born wildebeest within seven minutes can stagger to its feet and follow. Now the mother will lick it and proceed to recognize it as a being strange but her own. What, however, happens when herds are so dense that the unfortunate newborn after these minutes cannot recognize its mother? Let it make a wrong guess in the confusions of a wildebeest Times Square. The non-mother will butt it away. The mob will have swallowed the calf's identity. Lost, unprotected, it must in the end surrender its fate to the hyena or jackal.

The lion, preyed on by none, not too susceptible to disease or parasites, could in a few generations be a victim of overpopulation. The lioness produces her several cubs in a short period of gestation, and, should she lose them, comes into heat again at once. Yet in the Serengeti a stable population of about a thousand lions varies little in number from season to season. The area's immense numbers of prey animals, such as wildebeest, Thomson's gazelle, and zebra, could support far more lions. What keeps their numbers down? A subtle combination of behavior patterns, foremost among them maternal neglect, provides that only so many lions will reach a breeding age and situation.

The first control is territorial. As with the Australian magpie, only those females who are part of a permanently resident pride breed successfully. The second control is a dominance order like those of few other species. The young eat last. With Schaller I once watched a zebra kill where nine lionesses, rumbling at each other with the collective menace of a volcano, ringed the carcass flank to flank like daisy petals. In ninety minutes they ate some 450 pounds of meat, while a lone cub on the outskirts played with the zebra's tail, which he had somehow managed to secure. Had he sought a single bite before the lionesses were finished, quite probably he would have been killed.

This rank order of feeding, which places the males first though the lionesses have made the kill (and perhaps that is why our lionesses were eating in such a hurry, before males would show up), the females second, and the cubs last, compels no great deprivation so long as large game is available. A wildebeest or zebra will provide food for all. But the dry season brings a food shortage of a technical nature. The larger game migrates out of the grasslands into the woodlands. And the breeding prides, with their territorial attachment, refuse to follow. Through the dry season they live off the non-migrating animals, chiefly Thomson's gazelle—and a Tommie weighs at most about forty-five pounds. Adult appetite is seldom sated, and cubs go hungry. Should you in August glimpse two wan cubs waiting while their mother goes hunting, you may be fairly sure that only one will remain in November when the rains and the big game return. Yet food exists in plenty scarcely fifty miles away.

Infant mortality in the lion runs to about 50 percent. And we may contrast that with the quite opposite behavior of the adult hunting dog, which I earlier described, who will touch no food till the young are finished. One may deplore the lion, praise the hunting dog, but either judgment would be anthropomorphic. The behavioral contrasts are expressions of population control. The hardy lion, once he reaches maturity, will be around for a long time. Control must fall on the young. But the delicate hunting dog, so susceptible to disease, must do all in his power to keep adult social numbers replenished.

Another animal virtually immune to predation is the elephant. Non-territorial, the elephant in herds has historically migrated over immense African areas in search of food, water, and perhaps even change of scene. No major animal has been so little studied, and none, as I earlier described, exhibits such mysterious capacities of behavior. But the developing needs of developing peoples have taken away his freedom. Fields have spread. And the elephant, whose majestic presence once commanded the whole of Africa's majestic space, becomes today a prisoner of smaller and smaller protected areas.

Such an area is Uganda's Murchison Falls Park. Made famous by one of the world's most brain-numbing waterfalls, where the Victoria Nile passes through a rocky slot only nineteen feet wide, it has become even more famous for its elephants. They seem everywhere. Like a truck and trailer, a cow and her calf may thunder past your cabin door on the way to the camp's garbage cans. Once there was a great bull who acquired such a reputation as to become known as the Mayor of Murchison. Regularly he visited the tourists' cars in the parking lot, searching them for dainties. Visitors responded by leaving fruit in their cars, thus affording splendid snapshots to awe the folks back home. But the Mayor responded also. When he came to a car with no fruit, he turned it over. In the end, park authorities destroyed him.

A visit to Murchison Falls tempts the conclusion that the elephant is experiencing a population explosion. I was once so tempted, and I seem to have been wrong. Numerous they may be, but, in the opinion of Cambridge's Richard Laws, numbers are probably experiencing a slow decline. What the visitor witnesses is an elephant concentration camp. And the decline is the elephant's self-regulating mechanism responding to new conditions of density and restricted habitat.

The broad, crocodile-infested Victoria Nile separates Murchison's elephant herds into two distinct populations, each 7,000 or 8,000. The conditions on the South Bank are much the worse, both in crowding and destruction of natural habitat. An entire forest, its trees stripped of bark by the herds, stands today a leafless ghost. While in this well-watered land there is ample supply of grass, the elephant's traditional food, diminishing leafy browse may have brought on some nutritional deficiency. Or it may simply be the stress of density; we do not know. But through self-regulation the elephant is reducing his numbers to meet new conditions. The onset of fertility in the female is normally at the age of eleven. On the South Bank it is retarded to about eighteen. Calves are normally spaced about four years apart. On the South Bank spacing has reached nine.

The arrangements of lion and elephant illustrate the diversity of natural controls. But I present two rather different examples, ominous in human terms, derived not from any natural control but from human interference with natural arrangements.

That most sensitive of French zoologists, François Bourlière, has recorded two studies of deer that might be regarded as examples of the explosion-crash phenomenon were they not artificial. The first occurred on a large plateau in Arizona, just across the way from the Grand Canyon, where a stable population of 4,000 deer lived in balanced relationship with a fair number of wolves, pumas, and coyotes. The effect of predators on a prey population is almost always to weed out the sick, the malformed, the deficient. The net consequence, observed again and again, is to keep the prey population healthy. But early in this century human slaughter of the predators began and they were virtually eliminated. And with the slaughter the numbers of the deer began to rise. By 1920 there were 40,000, by 1924 over 100,-000. Then in a year it crashed to 40,000, and by 1939 was down to ten thousand. Over-grazing and food shortage had undoubtedly contributed to the peak crash, but it could not explain the final decline.

The other example came about through efforts of the United States government to build up a herd of reindeer as food supply for the local inhabitants of St. Paul's Island in the Aleutians. Here there were no predators at all to exert a selective pressure on the herd, and for many years the experiment seemed a huge success. It had begun in the autumn of 1911 when four bucks and twenty-one does were placed on St. Paul's. By 1832 they had increased in number to 500, by 1938 to well over 2,000. But then came what Bourlière well described as a cataclysmic decline. By 1950 eight remained.

While food shortage may have contributed to the Arizona collapse, it could have entered little into the shattering of the reindeer experiment. The only possible conclusion is that failure by any agency to remove the weak and the deficient from the breeding population gradually sapped the vitality of the whole gene pool.

It is a fate frequently predicted for the human species.

4

The innocent egg, like some accident-prone bystander, seems always to be getting itself into the impossible middle of arguments. When at the end of the nineteenth century Charles Darwin's notions about evolution were passing through their darkest days, he had many a witty antagonist. Among them was Samuel Butler, who gloomily remarked that, according to Darwin, a hen was nothing but an egg's way of producing another egg. Butler's classic line will live always, as we may assume that the egg's predicament is other than temporary. And we shall not wonder if the egg is today cooking hopelessly in one of ecology's steamier arguments.

The natural regulation of animal numbers is a phrase introduced to science in 1954 with the title of a book by David Lack, Britain's celebrated ornithologist. In that book Lack's concern for natural controls was generated by the variation of number of eggs, from season to season, appearing in a clutch. Then in

1962 the Scottish ecologist V. C. Wynne-Edwards published his *Animal Dispersion in Relation to Social Behavior,* which so deepened and broadened the concept that the sad little egg was left far behind. The ensuing argument between Lack and Wynne-Edwards has only begun its spread beyond ecology into the whole biological realm.

Simple records show the year-to-year variation of clutch size in many bird species. It is as if the bird possesses some peculiar picture of how many eggs should be in the nest. An American experimenter—who may or may not be described as a bird-lover —by taking away each newly laid egg once induced a bewildered but determined woodpecker to lay seventy-one eggs in seventy-three days. Lack, more compassionate, contented himself with such observations as egg-counting in a great-tit population while determining food supply. The young are fed on caterpillars. And in the particular years of his observations he found a startling correlation. The years of highest clutch size, eleven and twelve, were those when later in the season caterpillars appeared in greatest profusion. In 1951, when clutch size dropped to eight, caterpillars dropped to a minimum.

Granted that the observations were persuasive, granted that great-tit foresight plus family planning kept the collective appetites of the young within the bounds of what nature would provide, the question still remained, How did they do it? Lack met the argument with an answer not too convincing: that optimum clutch size would be determined by the highest number of surviving young, which in turn would be determined by food supply. But we may recall what happened to Elton and the sunspot theory. While the control of numbers might not impossibly, as in Laws's elephants, be influenced by the quantity or quality of food supply, still students going back to Malthus for answers had a bad way of getting into trouble. And Lack got into trouble. Many years later, in 1965, C. M. Perrins reported on the continuation of the great-tit studies which Lack had initiated. In a decade and a half of observation, the correlation of seasons appeared as coincidence. What emerged as significant was the relation of clutch size to population density. If the numbers of great tits double in an area, clutch size the next season may be expected to be reduced by two.

The bird-watchers should perhaps have paid closer attention to what was happening to the rodent-watchers. By now, however, all watchers were coming to a single conclusion: It is space, not food, that stimulates controlling mechanisms in most species. It is the condition of over-crowding, not the condition of under-eating, that holds down numbers. (And it is of course at this point of advancing theory that the human condition attracts compelling notice.) But mysteries remain. We cannot say that animals learn from experience to reduce the ambitions of the clutch. Even had they the capacity to learn, as Perrins pointed out, among great tits only half of the adults alive in one year will be alive the next.

It was in the course of the heatening argument between iconoclast and traditionalist that Wynne-Edwards published his mammoth, iconoclastic book; and I cannot believe that all those who attack him have read it. In that book he accepted the resources of the environment as the final limitation on animal numbers. But he denied that food supply, or competition for food supply, presents the direct influence on our numbers. He presented instead the hypothesis that evolution has favored *conventional* modes of competition which limit numbers well below the carrying capacity of the habitat. Male animals do not compete for such direct rewards as food or females. They compete for symbols, like territory or high rank in a hierarchy, which in turn present them with first access to food or females. The territory may or may not present the successful proprietor with an assured food supply. Most definitely no female waits on it, like a feathered Mae West suggesting that the male come and see her sometime. But the gaining of such symbolic prizes— limited in number—presents the successful competitor with prior access to food and females. The female is attracted to the prize, not to the proprietor. And since the numbers of such conventional rewards are limited, so the numbers of the population are restrained.

The lay reader may inspect such differences of opinion and wonder what all the shouting is about. But he must remember that the sciences are dominated by materialism; it is impossible for many a contemporary scientist to conceive of animals competing for other than such material gratifications as sex or food. Yet ethology demonstrates that animals rarely compete for either. The dismal consequence of such scientific obsolescence was sharply demonstrated at a recent Washington symposium by one of the most able of American zoologists, Harvard's E. O. Wilson.

The conference was sponsored by the Smithsonian Institution, titled *Man and Beast,* and enlisted many of our foremost scientific minds to consider various new hypotheses linking the behavior of man and other animals. I consider it compulsory to confess a degree of prejudice, since among the hypotheses greeted by no unanimous acclamation were those advanced by Konrad Lorenz and myself that aggressiveness and consequent competition are innate propensities in the human as well as other species. If exceptions exist, then they do not include *Homo sapiens.* Wilson was invited to address the conference on competitive and aggressive behavior.

His was a paper to leave the evolutionist's mind temporarily blacked out. For its premise he chose a 1939 definition of competition as "the active demand by two or more individuals of the same species (intraspecies competition) or members of two or more species at the same trophic level (interspecies competition) for a common resource or requirement that is actually or potentially limiting." Cordially he simplified the language: "Competition arises only when populations become crowded enough for a shortage to develop in one or more resources." It was a premise of Triassic obsolescence leading logically to the conclusion that competition is infrequent in the natural world. The conclusion lent biology's endorsement to environmentalism's tattered dream. At this widely publicized conference, Wilson's was the paper most frequently quoted in the sentimental American press.

It was the old story of what a false premise can lead to, even in the work of a most able mind. Wilson's definition of competition had entered the scientific literature just two years after Konrad Lorenz' first paper was translated into English. The science of ethology had yet to be born, let alone to receive a name. A generation of students of animal ways in a state of nature had yet to return from field and forest, their notes in hand. Their reports would be unanimous that while competition occurs everywhere in nature (can one speak of natural selection without it?) the competition within a species is almost never for material resources. Whether or not observers accepted all of Wynne-Edwards' hypothesis, or even his description of prizes as conventional, it could hardly be denied that the goals of almost all animal competition exist not in the material environment but in the behavioral environment of the animal population itself. A territory, for example, cannot exist in nature; it exists in the mind of the animal.

The controversy that has arisen around Wynne-Edwards' work expresses far more than the reaction of the unsophisticated to a highly sophisticated hypothesis. The opposition is to his concept of group selection, and it has enlisted as fastidious a scientist as David Lack. The fundamental problem facing any theorist attempting to relate clutch size or territorial competition or social rank to population control is how to explain the willingness of the individual to accept certain rules and regulations that exist not in his interest, but in the interest of the whole population.

The nesting habits of seabirds provide a clear illustration. A final limitation on the numbers of the kittiwake is offered not by food supply but by a scarcity of cliff-hanging nesting sites which the species favors as a defense against predators. The kittiwake pair can easily raise three young to maturity, yet three quarters of all nests in a colony will show an egg clutch of two. The population is thereby kept within reasonable bounds. But if you are a kittiwake and you have a nest, then why not three? What process of natural selection has induced this self-imposed birth control?

Early in his career, when Wynne-Edwards was working as a biologist in Canada, he took a superb photograph of a large gannet colony on a Newfoundland headland. The photograph, published in *Scientific American*, required the shortest of captions to tell its enigmatic story. One bump of the headland is white with breeding gannets on their nests. An adjacent headland is just as white with "unemployed" birds, those excluded from breeding because they have no nesting territories. But nests can be built anywhere. Why are only a limited number of sites acceptable as breeding stations? We may of course immediately skip to the consequence, that the arbitrary limit placed on nesting sites arbitrarily limits the population. But why do the unemployed birds, having competed and lost in their efforts to gain territories, accept the rules and regulations? They resemble nothing so much as human children in some game who have been declared "out" by the umpire and have been relegated to the sidelines. And we may ask, also, what processes of natural selection could evolve and enforce such a natural treaty between a population and its members?

To the discomfort of more conservative biologists, Wynne-Edwards proposed group selection, resting on the new principles of population genetics. It is the same answer which I have suggested for the stotting Tommie. In the whole gene pool of a population may evolve a behavioral trait of survival advantage to the group. A local population, in its capacity to surmount the hazards of its habitat for hundreds or thousands of generations, achieves a kind of immortality denied the individual. Any genetic change, whether or not of benefit to the short-lived members, if of benefit to the survival of the group should spread inevitably to become a part of the whole gene pool.

Let us think about Ellen Cullen's kittiwakes and presume that at some moment in kittiwake ancestry a mutation took place introducing a gene with a dominant allele limiting clutch size to two while the recessive remained the traditional three. The mutation would seem to be a disadvantage, since the parents would leave fewer offspring, and so should be eliminated. But would it not spread through the population with its problem of cliffside nesting sites? Group selection would say yes, And since the distribution of the gene in each generation is random, limitation of young lies beyond parental choice.

The significance of Wynne-Edwards' group selection—the superior survival prospects of the population with a superior gene pool—is that it provides a genetic explanation for the self-regulation of animal numbers. Such explanations are not easily come by. The traditional interpretation of natural selection in this century is differential reproduction, the proposition that superior individuals will leave more offspring to influence the succeeding generation's gene pool. But I suspect that the opposition to group selection goes back to a comment by Ernst Mayr that most biological controversies today are between those still thinking in terms of type and those who have moved on to think in terms of population.

Many biologists reject group selection as a concept unproved, others as one unnecessary, maintaining that the traditional interpretation of selection can explain everything. But the pathway of such explanations has been a tortured one, as we have seen in Lack's efforts to apply the traditional food theory. What seems at stake is less an explanation for a demonstrable phenomenon than a defense of an accepted if inconvenient definition. Perhaps the best comment was made long before the controversy became enflamed when M. E. Solomon reflected on the unreality of relating the fate of populations so exclusively to external environment. "The population functions in relation to a whole which includes itself."

The controversy will be resolved one day by the specialists involved. If Wilson was right at Washington, and competition occurs only when overcrowded numbers struggle for a scarce resource, then Malthus is confirmed. And humanity has little to look forward to but that chaotic day when in unlimited number we assassinate one another in our pursuit of inadequate resources. But if Wynne-Edwards is right, any population, human or non-human, has within its power the limitation of numbers through conventional rules and regulations and the capacity to abide by them.

While contemporary evidence seems to support Wynne-Edwards, there must always of course be those unlucky species who, lacking such powers and capacities, proceed on toward their unhappy rendezvous with decimation—in all probability, death by stress.

5

Readers of *Alice in Wonderland* have for generations accepted the March Hare in all his madness as a most ingratiating product of Lewis Carroll's imagination. That he exists, regularly and in number, must seem unlikely. But the snowshoe hare, when his time comes, dies off in the spring of an affliction resembling nothing so much as a nervous breakdown.

We have concerned ourselves with species that succeed in regulating their own numbers, and just how they do it. Now let us return to those cozy species that do not, and with a prayerful thought for our own species consider just what happens to them. I left the detective story when a clue was found indicating that population density, not food supply, was responsible for population crashes. And at an early date the clue suggested that Darwin might have been right, and that crowded populations could be the victims of disease spreading at epidemic rate.

It was about the time in the late 1930's when MacLulich was exploding the sunspot theory that a man named R. G. Green with various co-workers started picking up the corpses of snowshoe hares in Minnesota's Lake Alexander area. Few showed evidence of any infectious disease. Their manner of death, however, was odd. Some might be behaving normally, others might be in torpor, when suddenly they would be seized by convulsions and die. Another odd symptom was exhibited by hares captured, apparently healthy, and placed in captivity. The normal snowshoe hare tolerates the experience with indifference. But these in the springtime of the population crash died almost immediately. Autopsy showed a certain degeneration in the liver, a deficiency of blood sugar, and minor internal hemorrhages. Green described it as shock disease.

The description satisfied no one. Something specific enough to produce such similar deaths must be susceptible to definition. And why should it occur in the spring? Various hypotheses were advanced, the most persuasive by J. J. Christian, one of today's most earnest investigators. He saw the building up of a population as a time of intensifying stress. The increasing number of young, the increasing competition of adults, the increasing number of strangers in a massive, increasingly disorganized population at last brings on a state of exhaustion both psychological and physiological. It is as if the cycle's last winter with its normal hardships sets the stage for the entrance of the last straw. And that last straw comes with the sexual demands of the spring. Everybody drops dead.

Whether or not the hypothesis is correct remains still unproved. The notion that the mad March hare perishes at last from love, romantic though it may be, comes from one of our most skillful endocrinologists. More significant, however, than the disagreeable fate of the snowshoe hare was the introduction by such students as J. J. Christian and David E. Davis of the physiological consequences of social stress in high-density populations. Further evidence from the field emphasized the lethal relationship. Most studies had been made of rodents, particularly susceptible to population crashes. But the shrew, studied in Manitoba tamarack bogs, is an insectivore like the mole. During a population explosion in 1957 the excitability of animals was such that even early in the eruption they lived only four days in captivity, though, like the hare, they offer no normal captive problem. By autumn, and the peak of the explosion, they lived only eight hours. By then the average number of embryos carried by a pregnant female had decreased from seven to three. But the uncommon contribution of the shrew study was what happened to space. Population increased from less than one per acre to ten. And in the masked shrew, the species most affected by the explosion and crash, there was no increase in overlapping of territories. They remained defended, and shrank and shrank with ever more pressure on proprietors.

The increasing stress placed on an exploding population could not be better illustrated. And now the attention of science, with ample field material available, was shifting to the laboratory. And a single inspired experiment, confirmed and reconfirmed, has been enough to finish the notion that birth control violates natural law.

The house mouse is territorial, and the female under conditions of normal density encounters no strange males. The experiment, which exhibited what later became known as the "Bruce effect," was first conducted in Britain. A female mouse was impregnated by a stud male. If within four days she was mounted by a strange male, she aborted. The implication was that of a morality in mice previously unsuspected. But the investigation went further. If the impregnated female even *saw* a strange male within four days, the chances were fifty-fifty that

pregnancy would be terminated. The final experiment—of utmost significance to population dynamics—demonstrated that the same failure of pregnancy would come about if she were placed in a cage where a strange male had been and she merely smelled his recent presence.

The experiment had been conducted under most elegant conditions, with ample numbers of subjects and ample controls who, sniffing no strange males, proceeded on with their normal pregnancies. Even so, stunned biologists suspected that something must be wrong and set up new experiments. Bruce had used an albino strain of laboratory mice, and such inbred creatures frequently yield untrustworthy conclusions. And so another investigator set up the same experiment with wild deermice. The results were the same. Physiologists demonstrated the cause: It is the smell, in all situations. The odor of the strange male frees the fertilized egg from implantation.

A last experiment performed at Oxford, and published in 1968, has shown what a powerful mechanism is the Bruce effect, and how widespread it must be in the control of animal numbers. Mice are known as spontaneous ovulators; that is, like most mammals, the female ovulates at regular intervals when fertilization must take place. But there are animals like the vole who are induced ovulators. Copulation releases the ovum to descend and meet the sperm already waiting. And so twenty female meadow voles were mated and permitted to remain with their males. Sixteen produced litters. But another twenty, after mating, were exposed to strange males. Only five produced litters.

The Bruce effect affirms natural birth control in its purest form. An overcrowded population, introducing strange males to the vicinity of a pregnant female, produces immediate abortion. In all probability, comparable effects of which we are yet unaware explain in many species, perhaps even the elephant, the reduction of embryos. But the simple effect of stress due to density cannot alone be responsible for the control of numbers.

A frequent observation has been the variation of response to growing density by different groups of quite the same creature, living under quite the same conditions. While all at some point must reach a point of reproductive breakdown, levels of tolerance may differ widely. It is as if the populations differed "temperamentally," to use Charles Southwick's word. Aggression may increase rapidly or slowly. Mother-infant relations may disintegrate with little increase of stress, or remain respectable at a much higher level of crowding. And the difference will probably be due to the presence or absence of a very strong alpha in a given group.

The relation of rank to stress has its grim side. The overdominated animal may with small ado lie down and die. It has happened to cockroaches that the badly beaten animal, unwounded, has died apparently of nothing but discouragement. Rats introduced to established groups suffer persecution and may die within days. In a Glasgow laboratory a rat died after ninety minutes of persecution. He had no significant wound nor had he suffered the least internal injury. He died of stress.

We believe that subordinated animals experience enlargement of the adrenal gland, and under the pressure of sufficient stress may, through adrenal exhaustion, sink into apathy or death. But a curious quality of the alpha is his relative invulnerability. We know that the true alpha male is born, not made, and perhaps it is a portion of his genetic distinction that he endures stressful situations with a glandular equanimity greater than others. The same may be true of the alpha female in those species where female rank orders exist. In Australian experiments Myers has shown that among rabbits subjected to density pressures it is the low-ranking female who suffers the greatest fetus mortality. We may speculate, then, that the "temperament" of a population may well be determined by the random incidence or absence of a powerful alpha, male or female, whose very presence acts to forestall the disintegration of social organization.

The relative immunity of the alpha and vulnerability of the omega was reported as early as 1952 by John Calhoun in his Maryland experiment with rats, and is suggested as late as 1968 in a study of men. In that year our American journal *Science* published a medical study of all 270,000 male employees of the Bell System Operating Companies, the telephone organization that in the United States is a near-monopoly. The mammoth investigation linked educational background, job achievement, and incidence of coronary heart disease. No credo could be more widely accepted than the belief that it is the striving, the intense competition, and the responsibilities of high achievement that in American life curtail men's lives. No credo, as it turned out, could be more false.

The Bell System offers, like a perfectly arranged laboratory condition, a single controlled environment. Operating units whether in Georgia or New York State have similar structures, fulfill similar functions, provide similar jobs. All is directed by a single top-management policy with the same system of pensions and security, insurance and medical practices, and, perhaps most important, record-keeping. And the 270,000 histories provide a sample so large that even small variations from the expectable would have statistical significance. The variations were not small.

Reading from bottom to top in the telephone company's pecking order, we find that workmen contract coronaries at the rate of 4.33 per thousand per year. Their immediate superiors, the foremen, have it slightly worse, 4.52. But supervisors and local area managers drop to 3.91. Then comes a bigger drop. General area managers have a mere 2.85. We then come to the high competitors, the high achievers, the high executives. Coronaries occur at a rate of 1.85, about 40 percent of the level of workmen. And while we may say that many a coronary customer could have been eliminated before reaching the alpha rank, we must also reckon that the high executives are older.

There will be no problem of interpretation for those schooled in the population dynamics of animal groups. Certain environmental influences undoubtedly made a contribution. The study revealed that college men are a far better risk than non-college and one may fairly suspect the influence or better environmental backgrounds. But the investigators pointed out that the single worst record was made by college men who rose no higher than foremen, while non-college men who rose to the top shared the relative immunity of their fellow executives. The report ended up puzzled, but admitting that something biological must be going on.

Something biological was most distinctly going on. If man is an evolved animal, then the natural histories of alpha monkeys, of alpha rabbits, of alpha antelopes, of alpha fish must provide some hints about the natural endowments and deficiencies of the alphas and omegas among us.

6

Had Malthus been right, and could food supply be demonstrated as the literal limitation on animal numbers, then we might reasonably interpret contraception as a violation of natural law. But in strange concord both Malthus and Paul VI were wrong. Throughout animal species self-regulatory mechanisms provide that population numbers will never challenge the normal carrying capacity of the physical environment.

It is an error to draw a clean line between the learned and the innate. Quite obviously the lion cub through unpleasant experience must learn to obey a law of subordinance which in itself is an innate social disposition in the species. Quite obviously the herring-gull chick, in one of Tinbergen's studies, must learn the territorial boundaries of its parents if it is not to stray too far and get pecked to death. Learned elements enter into almost all animal arrangements, and so to varying degree we may regard them as sub-cultural, just as Wynne-Edwards' conventional prizes are in part innate, in part traditional and learned. But in that most cultural of species, man, for whom tradition has so widely replaced patterns of innate behavioral or physiological command, contraception has become a cultural substitute for behavior that would otherwise have come so naturally.

We may regard it as a pity, perhaps, that our young females, unlike the lapwing, are sexually responsive to unpropertied males. We may sigh that our omega males, unlike the Uganda kob, do not cheerfully accept psychological castration. We may regard it as deplorable that in our family arrangements we do not, like the great-tit, respond to rising population by laying fewer eggs or, like the house mouse, spontaneously abort. We may with less certainty look askance at the lemming's youth movements, under the stress of intolerable numbers, conducting suicidal marches, or at the snowshoe hare dropping dead; for we may just possibly resort to such lugubrious impulses ourselves one day.

However we may regard in human terms the gradual loss of such innate mechanisms, we cannot blame that loss with entire conviction on the mid-Pleistocene expansion of the human brain. Without any doubt the rapid enlargement of our cortical equipment exerted increasing inhibition on old forms of compulsive behavior. Even so, in our more primitive days we stayed with natural law and substituted social traditions for previously natural processes.

The American ecologist J. B. Birdsell has shown that by natural increase of numbers the aborigine would have reached the food limitations of Australia in two thousand years. But he was there far longer, and he never came close. Disease, territorial spacing of groups, tribal warfare may have made contributions. But the principal factor of population control was infanticide. The evidence is as conclusive in the Eskimo, observed before modern influences had modified his ways, as in the Australian aborigine. Both were hunting peoples with a pressing need for active males. In both, the proportion of young males to young females was approximately 150 to 100. Girl babies had been the chief object of destruction.

When in 1922 Carr-Saunders considered the population problem, he was unaware of the long evolutionary history of the animal control of numbers. His concern was with primitive peoples, and his assumption was that population control had begun in the Stone Age. His comprehensive review of almost all then known about primitive peoples led to his thesis of "optimum numbers." Within every group there is a number well under any threat by starvation, yet sufficient to gain a maximum yield from the environment. The number is sustained by varying traditions—by infanticide or compulsory abortion, by cannibalism, head-hunting, human sacrifice, ritual murder, by tabus against incest, or against intercourse during the period of lactation. The environment is held to a constant size either through outright territorial defense or through traditional attachment to a familiar area, formalized as in certain aborigine tribes by belief that the home grounds are inhabited by the spirits of dead ancestors. Out of his conclusions concerning optimum numbers, and anticipating Wynne-Edwards' concept of group selection, Carr-Saunders wrote:

> Those groups practicing the most advantageous customs will have an advantage in the constant struggle between adjacent groups over those that practice less advantageous customs. Few customs can be more advantageous than those which limit a group to the desirable number.

With the coming ten thousand years ago of the agricultural revolution came a wrench to all those traditional forces that has severely limited numbers in the ancient hunting bands. Numbers of people were in demand. Yet even in a modern farming tribe one finds customs specifically regulating sexual and reproductive behavior. Jomo Kenyatta needs no introduction to the contemporary citizen. Few, however, are aware that his *Facing Mount Kenya*, written when he was a student at London University, is one of the most perfect monographs in the literature of ethnology. As a westernized mind, he brought scientific discipline to his subject. But as that rare anthropologist, a member of the tribe he describes, he brought the intimacy of birth and young experience to his analysis. He confirms population dispersal through territoriality with his comments on land tenure: that while the Kikuyu defended their country collectively, "the fact remained that every inch of the Kikuyu territory had its owner, with the boundary properly fixed and everyone respecting his neighbor's." He considers customs of division of labor, and of education. But nowhere is there a passage more fascinating than his description of those traditions resolving adolescent sexual drives with limitation of young.

At the rites of puberty the male is circumcised, and the female subjected to clitoridectomy, thus reducing her capacity for sexual excitement. Until puberty, masturbation is accepted as a normal boyish practice, and, while indecent in the presence of elders, within an age group it may even be a subject for competition. After initiation, however, masturbation is regarded as babyish. Now the custom of *ngwecko* takes over, a restricted form of intercourse which Kenyatta describes "as a sacred act, and one which must be done in a systematic, well-organized manner." *Ngwecko* takes place in a special hut, the *thingira*, a rendezvous for the boys of an age group which girls may visit at any time. Here they eat and drink, and if the boys are in the majority, then the girl may choose her companion, although it is considered selfish for a girl always to choose her most intimate friends. When partners have been arranged, the boy strips, but the girl retains her *mwengo*, a soft little leather apron, which she pulls carefully between her thighs. Normally during the *ngwecko* the partners experience sexual relief, though since the girl has been circumcised she may not have equal necessity. What is quite tabu is disturbance of her apron. Any accident of conception is severely punished by tribal law, and if in later years the girl approaching marriage proves not to be a virgin, then her value is drastically reduced.

From Kenyatta's account we may see how finely balanced are those Kikuyu traditions which, accepting human sexuality, reduce the consequence of unwanted young. Yet with the breakdown of tribal discipline in our time must come the breakdown of such customs and a cultural degeneration leading to the population explosion.

In a sense it was not the eighteenth century's industrial revolution but the simultaneous cultural triumph of humanism that most effectively destroyed the older cultural institutions which, with small sentimentality, had replaced the still earlier biological institutions. Humanism's respect for the dignity of man, and its regard for every human life as sacred, while among the most powerful forces ever to advance the welfare of men along certain fronts, had ambiguous consequences on others.

Throughout a large world of primitive societies missionaries and colonial masters reacted with equivalent horror to such institutions as cannibalism, head-hunting, and human sacrifice. Tribal warfare, particularly in Africa, was put down. Infanticide was discouraged, at least within the limits of administrative process and missionary persuasion. While the destruction of twins, a custom throughout all of black Africa which effectively reduces the breeding population by 2.5 percent, has through parental cunning evaded authority even to this day; still almost all tribal ways affronting the humanist standard were reduced but not replaced. When I found myself in Kenya in 1955 a terrified observer of the Mau Mau rising, I came reluctantly to accept a Kikuyu fact of life. While all the world was aware of Kikuyu grievance concerning land, quite unreported—perhaps because it was news unfit to print—was Kikuyu anger at British efforts to suppress clitoridectomy.

Perhaps humanism's supreme triumph has been its transfer of guilt as a sense of sin against God to a sense of sin against fellow man. But again the consequences have been ambiguous. With the advent of modern medicine and biochemistry our regard for every human life as sacred received most miraculous tools. The rate of infant mortality dropped like a rock in a well. Life was so prolonged that a new class of senior citizen came into being. Breeding populations in the more advanced countries were now little reduced by the death of young mothers in childbirth. Strangely enough, with modern nutrition in the same countries the onset of menstruation and fecundity dropped three years in less than a century. Not at all strangely, however, the abortionist was condemned to surgery's red-light district. And driven on by our sacred mission, through mass-produced drugs and insecticides we extended the new breeding potential to all peoples, advanced or otherwise.

Religions are rarely humane, and certainly thus far humanism has been no exception. The guilt of the humanist in his preoccupation with the numbers game has been his sacrifice of human quality for human quantity. Life must be prolonged, whatever agony it presents to the dying. A child defective physically or mentally must somehow be saved sufficiently to join the breeding population. To restrict the reproductive rights of the genetically afflicted would be an act of discrimination. Not impossibly a sick society is one preoccupied by the fate of the sick, not the well.

We shall find out one day if, as many biologists fear, overprotection of the human being, like underpredation in the reindeer herd on St. Paul's Island, will produce a genetic collapse in those very populations most addicted to the compassionate way. If so, it will be an appropriate biological conclusion to a misapplied philosophy. Yet I find it a conclusion too appropriate. Human arrangements are seldom so tidy.

We should look for simpler answers. There is sufficient evidence, I believe, to warrant the conclusion that Malthus was wrong and that human numbers will never reach such magnitude as to encounter the limitations of food supply. Long before such a rendezvous can take place other forces will have affected our numbers. And if we take nature as a model, then the maximum probabilities are only two. The first is birth control.

There is sufficient evidence also, I believe, to warrant the conclusion that the papal decision should not have condemned contraception as a violation of natural law. Birth control, whatever form it takes, is a cultural substitute for biological mechanisms prevalent in the natural world. As our population problem has a cultural cause, so we are provided with a cultural answer. But that answer must be mandatory. I have suggested that as a portion of the social contract the individual has no right to expect society to provide equal opportunity for more young than the group can handle. We have seen that in animal species the numbers of young are not determined by parental choice. And so we must look for means of enforced contraception, whether through taxation on surplus children, or through more severe but more democratic means such as a conception license replacing or supplementing the marriage license. Abortion should be freely available to those suffering unintended pregnancy. In international relations, of course, any material aid to peoples who through ignorance, prejudice, or political hypnosis fail to control their numbers should, I believe, be forbidden.

The program sounds more formidable than it would prove in practice. The vast majority among us accept traffic regulations without resentment because we are aware of their necessity. Conscience too has a way of internalizing what has started with external pressure, and of transferring to the voluntary what was once the compulsory. Even fashion makes its contribution, so that what is socially unacceptable becomes something that is simply not done. Most hopeful of all is the demonstrable proposition that a cultural institution, such as private property, which accords with natural law rarely fails.

We have of course an alternative to a solution both sane and humane. But the alternative, death by stress, is a messy one indeed. If we recognize that population density, not food supply, is the chief factor limiting animal numbers, and if we recognize also that no population increases indefinitely, then, however unattractive the alternative control may be, we must come to accept it, even to applaud it. Such a program of madness is available.

The rising rate of automobile accidents is a quite perfect example of a form of population control mathematically determined by population density. Granted that we have accepted an insane solution as preferable to a sane one, then we must see that the automobile, which strikes most heavily at the young, is indeed an excellent instrument for reducing the breeding population.

Another agency striking hard at the young is drug addiction. While we cannot as yet be sure that drugs reduce reproductive potential, still we should be wise to gamble that they do, and to offer every encouragement to widespread addiction in the young.

The trouble with cardiac and other stress diseases is that they tend to reduce numbers in those who have passed the breeding age. Even so, we know that the omega is far more susceptible than the alpha. We should therefore encourage in business, for example, all those tendencies toward mergers and ever magnifying organizations that reduce in number the immune alpha and infinitely expand the ranks of the omega.

If life in the megalopolis discourages large families, then perhaps by discriminatory taxation falling most heavily on real estate we might reverse the flight to the suburbs and drive middle-class families back to the unpleasantness of urban reproduction. We should have nothing to lose by such a move, in any case, since further concentrations of city life must produce more stress, more broken marriages, more impotency due to acute alcoholism, more corpses, victims of crime in the streets, and more couples living together in unreproductive sin.

Homosexuality should not be neglected. Already it subtracts 4 or 5 percent from the American breeding population. We could do better.

Suicide too offers splendid vistas. We fall far behind such advanced peoples as the Swedes and the Swiss. Here too we could do better. Student suicide is rapidly increasing. But as stress and density close in upon us I feel a confidence—or perhaps it is no more than patriotism—that we Americans will catch up, and that suicide particularly in the young and the discouraged will make a significant reduction in our breeding numbers.

What at all times we must keep brightly in mind as, like thoughtfully mad March hares, we inspect the real possibilities of death by stress, is that while any reduction in numbers is a gain, significant reduction can only be accomplished in the young breeding group. The reader may wonder, with such an admonition, why I have so ignored war as an instrument. But war, in my opinion, has seen its best days. Its growing unpopularity with those who must fight it may turn out to be a passing whim. More serious is war's increasing preoccupation with the wastage of expensive machinery rather than with the traditional wastage of inexpensive men. Wars simply do not kill enough people. A nuclear entertainment would of course leave us with no population problem at all. But even as highly publicized a war as that in Vietnam has failed throughout its entire course to kill as many Americans as that magnificent engine of destruction, the automobile, kills in a normal year.

We must look to more imaginative agencies than war to dispose of the immense numbers who must someday die of stress. And I am sure that some future survey of likely instruments will reveal lethal conclusions of a wonder that the imagination cannot glimpse today. When our population has again doubled, when not a water supply remains unpolluted, when the traffic jams of tomorrow make today's seem memories of the open road, when civil disorder has permanently replaced war as a form of organized violence, when the air of the city can no longer be breathed and the countryside has vanished, when crime has become such that no citizen goes unarmed, when indigestion becomes a meal's final course and varieties of rage and frustration remain the only emotions man or woman can know, then perhaps, if we are young, we shall comprehend the lemming.

But of course future times of such stressful wonder may never come to be. Somewhere along the road we may have chosen compulsory contraception. And yet no one can make a sure prediction. *Homo sapiens*, that creature mad beyond the craziest of hares, lunatic beyond all lemmings, may go to the end of the road with no impulse more logical than to discover what lies there. How high is the mountain, how profound the stream? Which in the end will bend the ultimately defeated knee, we or our world? Shall we embrace the logic of limited numbers, acceptable to elephants and mice, or shall we mount the hilltop and defy the winds?

One cannot say. The tragedy and the magnificence of *Homo sapiens* together rise from the same smoky truth that we alone among animal species refuse to listen to reason.

7. *Space and the Citizen*

The ten-spined stickleback, as a fairly simple exercise in arithmetic will reveal, is a fish with seven more spines than a three-
spined stickleback. I have sometimes suspected that when
Desmond Morris, the famous author of *The Naked Ape* and
The Human Zoo, was a student at Oxford almost twenty years
ago and chose as a principal object of study the ten-spined stickleback, he in truth was performing a delicate exercise in one-

upmanship. His maestro, Niko Tinbergen, had devoted an immoderate portion of his career to study of the three-spined stickleback. By choosing the ten-spined, Morris with the simplest of ploys left Tinbergen seven spines behind. Then, when his work was published, the young zoologist whose genius for titles would one day be proved, legitimized his claim to the far-out with one of the farthest-out titles in zoology's literature, *Homosexuality in the Ten-Spined Stickleback.*

In an early season of my own research I encountered a reference to Desmond Morris' paper in somebody's bibliography, and I could scarcely wait for my next visit to London and the libraries of the British Museum to discover for myself what had been going on at Oxford. And much indeed had been going on in Tinbergen's dignified department. What Morris had begun as an experiment in overcrowding and its relation to aggression had concluded with observations that no right-thinking scientist could have anticipated.

Aside from their spinyness, the two species have a few distinct approaches to reproduction. The three-spined digs a nest in the sandy floor of some shallow backwater. Through the irresistible appeal of his red belly, he will attract a female to his nest and there she will lay her eggs for him to fertilize. But the ten-spined will have nothing to do with so crude a home for his children. In similar shallow waters he finds patches of waterweeds and among them, using the weeds for material, he builds a spherical nest pierced by a tunnel. Surrounding it is his territory, and when the breeding season starts, there is, as in all sticklebacks, an uproar of fighting with other males. But as is customary in territorial behavior, when borders are defined neighbors settle down to tending their nests and ogling females. The trouble-making, ten-spiked-devoted scientist, however, had purposely chosen an experimental tank large enough for only two territories, in which he generously placed five males. Three were necessarily beaten by their betters and retired in frustration to the corners of the tank. The winners, having built their nests, now waited for proper companions.

The sex life of the ten-spined stickleback has its lurid aspects, even when all goes well. If a female ripe with eggs swims through a territory, the proprietor will attack her as he attacks all intruders. But if she is ready enough, then she will not flee, and if her belly is swollen sufficiently to inflame his interest, he will do a little courtship dance. The dance is a succession of little ducking jumps enticing her to his nest. At its entrance, timidity may overcome her, in which case he will bite her. Bitten, she will either flee and he will start looking for a more responsive female, or she will be overwhelmed by his attention and enter the tunnel of his nest, her tail protruding. Now he arranges himself beside her tail and shivers his nose against it. The excitement of his shivering nose being altogether too much for her, she will lay her eggs in the nest and move on. Now he enters the tunnel, fertilizing the eggs as he passes through. Emerging on the far side, he drives her away. Not only her fun but her maternal chores are finished; he will raise the young.

Such are normal sexual relations in ten-spined-stickleback life. But Morris had created an abnormal situation in the crowded tank with its subordinated, surplus males. And so, when he spied an omega male stealing through the weeds into the heart of a territory while the proprietor was off charming a female, he assumed that the fish was trying to steal the nest. But this was not the ambition. When the female fled, the subordinated male immediately took the position where she had been. The proprietor turned from his dance and, perhaps blinded by lust, failed to recognize that he was now facing a pretender. He did his courtship dance again, and now the pseudo-female followed, just as a female would do. His nose close to the proprietor's two little white ventral spines that become erect in courtship, he was led to the nest. A true female might hesitate and require a bite. The homosexual required none. He entered the tunnel, leaving the proper amount of tail protruding. The gulled proprietor shivered his nose against it. The homosexual accepted the attention for precisely the length of time that a female would have taken to lay her eggs, about forty seconds, and then fled. Now it was the potential father's turn to be frustrated, for of course on entering the nest he found no eggs.

The strange incident seems to have been no matter of aberrant ten-spined personality. Morris observed it again and again. On one occasion all three of the surplus, dominated males tried simultaneously to receive the sexual attentions of the thoroughly addled proprietor. While later laboratory experiments with overcrowding might yield more drastic results, that of Desmond Morris back in 1952 remains the most bizarre.

Among the more ominous experiments, that made by John Calhoun at Washington's National Institute of Mental Health has received widest attention. It was at the beginning of his career that Calhoun demonstrated the natural regulation of numbers in a rat population, as I earlier described it. Now his purpose was to force overcrowding. The experiment was conducted indoors, under controlled conditions, and its chief ingenuity was an arrangement of four interconnecting pens. The two end pens had each but one entrance which could be guarded by a strong male rat. But the middle pens, each with two entrances, could not be so guarded and so became the center of free social action. As the population rose, the action in these middle pens became such an animal nightmare that Calhoun was to describe it as a "behavioral sink."

In each pen was an abundance of food, water, and nesting materials, along with artificial burrows for nesting mothers. And each was large enough to accommmodate comfortably a group of twelve adult rats, the number that Calhoun had earlier learned was a normal size in a natural population. But as numbers rose and crowding took place, nothing like an even distribution of numbers occurred in the pens, just as nothing like equivalent behavior developed.

At the beginning of the experiment a natural status struggle took place among the males, and each of the most powerful alphas to emerge took possession of an end pen. Since there was but one entrance, each had only to take his position beside it to guard the pen and keep order within it. Here his harem of females made their nests undisturbed. Calhoun wrote, "In essence the male established his territorial dominance and his control over a harem of females not by driving out other males but preventing their return." He slept by the entrance, paying no attention to the comings and goings of his females. If a male appeared, he was instantly alert, but he would tolerate the entrance of one who accepted his dominance. Such subordinates frequently slept in the pen. They never interfered with the females or attempted copulation. In general the harem females made good mothers, built proper nests, nursed and protected their young, and raised about half to weaning. But if there was order in the end pens, there was chaos in the middle.

As population rose, its weight fell on the unprotected middle pens. Here there was a class of dominant males, but they could hold no territories, and fighting was frequent. Rank order shifted with victories and defeats. There emerged also a middle class frequently attacked by the alphas but rarely contending. They were sexually active, but seemed unable to discriminate between estrous and non-estrous females, or even between females and juveniles. And then there were the totally subordinated omegas who "moved through the community like somnambulists," who ignored all others and were ignored by all. To look at them, they seemed sometimes the fattest and sleekest, and they remained unscarred by combat. These were the drop-outs.

So far, such a class structure in a crowded population might be expected. But with further crowding a class quite beyond prediction emerged. Calhoun called them "the probers." They were subordinate but active, and formed an essentially criminal class. They moved sometimes in gangs. Attacked by a dominant, they never contended, yet were never discouraged. They took no part at all in the status struggle, but they were the most active males in the colony. Hypersexual, they were also homosexual, and frequently cannibalistic too. Most astonishing was their tendency toward an action which can only be called rat rape. As the stickleback has a courtship ritual, so has the normal male rat. If he pursues a female to her burrow, he waits outside for her to emerge and accept. He never enters. But the probers dispensed with all courtship, chased females into burrows, trampling and killing young with abandon. Following their sexual activity, if dead babies lay about, they ate them.

The disorder of the middle pens disrupted female life more than male. Nest-building suffered first, and throughout the last half of the experiment no nests were built at all. Infants were scattered, abandoned, eaten. Whereas infant mortality in the protected end pens was held to 50 percent, in the middle pens it was 96 percent, almost total. Half of the females themselves died of pregnancy disturbance or sexual assault.

Calhoun's experiment, with all its horrifying human impli-

cations, caused wide comment. Yet one most incisive observation has been little recorded. The rats enjoyed the behavioral sink. The females in particular seemed incapable of resisting the social excitement of the middle pens. Protected harem females sought the middle pens when in estrus, without interference from their overlords. Although food and water was available in the protected pens, almost all sought the middle-pen hoppers. Their behavior in the rat mob was no different from that of the non-harem females. The one advantage protecting their young was the peace and the order of the end pens, to which they could retreat. But while the end-pen alphas never left their estates, only 3 percent of their females could resist the crowd.

People and times may change. But any consideration of space and the citizen must recognize the historic truth that people have tended to cluster in cities because they wanted to.

2

The evolutionist must approach the encounters of urban men with caution: the city is a human invention. He may accept the challenge of population control with quite a different order of confidence, since animal precedent demonstrates the self-regulation of numbers as approximating natural law. Conclusions become possible. He may approach such a painful exhibition as the revolt of the young with the assurance that since the young of all species have had their problems, an evolutionary approach may throw a fresh spotlight on shadowed areas previously unrevealed. He may delve into the question of leadership, and the alpha fish, and just what is such a fish made of, with the secure understanding that in social species examples of the alpha and of the status struggle have never been lacking. In none of these areas of contemporary dilemma are we bereft of animal example. They are situations common to natural arrangements, and from nature we may hope to extract information, hints, admonitions, sometimes laws. But the city was made by man.

For long we approached the city. We came to worship at lost temples impelled by lost dedications. We came to enjoy the excitements and exchanges of ancient markets, to beg or to bargain, to find pleasures in the stimulation of the throng. We traveled long distances sometimes. In a city of the Turkish highlands many thousands of years old there was even a foreign quarter, where envoys and traders from the distant Euphrates customarily stayed while they dickered for such treasures as obsidian. Old though it may be, the city as we know it was invented by man. And while we may look to the prairie-dog town, and discover in its ordered, competing yet integrated neighborhoods a flash of suggestion for urban man, still it is far analogy: the prairie-dog town is no Paris. Or we may look to the arrangements of the Japanese monkey with colonies larger than any known in the primate world: yet this is no Tokyo, no city such as you and I must know and endure. We face in the urban concentration something new under the sun, something unanticipated except by the biologically, genetically directed termitary: but we lack the insect's genetic directives as we lack an evolutionary common ground. While we may live in our cities like ants in an ant-hill, as vertebrates we are genetically unprepared for such contingency.

Urban concentration must be approached with utmost care by the evolutionist. It is a challenge of a central sort in the contemporary condition. In that grand tradition of the modern human being, we oppose nature, master our environment, and, victims of nature's tricks, we are threatened by the urban environment which we in our hubris have created. The evolutionist's comments on the city must be limited to his comments on man.

Let us write down one more extension of the social contract: The group must present to all its unequal members equal opportunity to develop their genetic potential; in return the individual must by coercion or consent sacrifice any right to produce young in greater number than the society can tolerate, or the fulfillment of the individual must be suffocated by indiscriminable numbers. So, following the contract's equity, we proceed to a guarantee that must come from society: spatial arrangements must be such that minimum distortion, psychological or physiological, may inhibit or divert the development of society's members. But what are these arrangements? And how much does overcrowding as such distort us?

Ambivalence of animal attitudes toward space complicates an investigation. We seek space, but we seek also to destroy it.

Calhoun's rats chose the excitements of the middle pens; had peace and distance been all they sought, no behavioral sink would have occurred. And we may say of urban problems that none will ever be solved by men willingly turning their backs on urban life. Human solutions must be sought within the context of innate needs, and overcrowding itself must be inspected with care.

Our preoccupation with population dynamics is producing a new scientific romanticism which explains all human fault by overcrowding. Such a fashionable view is presented by the Russells' recent *Violence, Monkeys, and Men*. The book is badly marred by the Russells' failure to acknowledge that men and other primates have led separate courses of evolution for at least twenty million years. The selective pressures that we have survived have been of quite different order from those of the ape. The book is likewise marred by a dismissal to a short appendix of the enormous question of man's hunting past, and what contribution to our violent present was made by the necessities of our evolving way. Their statement that only in the last fifty thousand years has man "lived to a large extent by hunting" is one for which they fortunately make no attempt to provide authority, since no authority exists. Nevertheless, through careful selection of evidence that violence occurs only as a consequence of overcrowding, and through just as careful avoidance of contrary evidence, they present a persuasive case.

An abundance of studies may be drawn upon to illustrate the violent consequence brought about by high-population density. Why the Russells did not compare three different studies of the langur in India, I do not know, since the evidence is so superficially convincing. The langur is a leaf-eating monkey, living in troops of twenty-five or so. Distributed widely in India, its conditions of life vary as widely; and it is the variation of population density that illuminates the comparative studies.

The first observations of the langur were made by Phyllis Jay in certain forests of central India where the creatures are fairly scarce. Each troop has a range of about two square miles and rarely contacts another. While troops tend to avoid one another, there is no conflict if contact occurs. There are no defended territories, and no evident boundaries between groups. Within the little society males have a rigid rank order, and almost never quarrel. The nearest thing to aggression may occur when a dominant male, copulating, is surrounded by a group of heckling adolescents.

The langurs of India's central forests seemed the ideal, sunny, non-aggressive creatures of legend, and Jay's study, completed in the early year of 1959, did much to reinforce the arguments of those primate students that monkeys never fight, never defend territory, never do anything but behave themselves in a fashion rarely glimpsed in human schoolyards. It was a time when we all still said that "langurs are this way."

Then, however, Suzanne Ripley made her equally careful study of langurs in Ceylon. Troops were of about the same size. But nowhere did there exist those infinite distances for the happy, wandering life. The troop's two square miles of India's central forests became an eighth of a square mile in Ceylon. And here there were not only territories, with actively defended, unchanging borders; groups sought combat. Like the howler and the callicebus, the langur is a noisy monkey. Morning treetop whoops would bring defiant answers from whooping neighbors and mobilization on the border. Ritualized displays might take place, with vast leaps through the trees. But in these combats between groups true fighting could take place too, with chasing, wrestling, biting, tail-pulling.

It is important to note that for these leaf-eating monkeys no shortage of food inspired conflict. Even the most arid zones of forest could carry numbers well beyond the actual population. Neither was any great damage inflicted by the continual conflict. Despite what must be described as violent behavior, in the best territorial tradition a maximum of excitement was generated with a minimum of physical harm. And, finally, the aggression directed outward infected not at all relations within the group. Although all groups studied by Ripley contained at least two adult males, and one contained six, harmony was near-perfect. Even females joined with their males in the territorial battles. This record of inner harmony was not to be duplicated in the third study.

Yukimara Sugiyama, of Kyoto University, went to India to gain some perspective on observations of the Japanese monkey. He found his perspective. Perhaps by chance Sugiyama chose the langur as a species and the Dharwar forest in western India as his scene. And there he encountered the nearest equivalent to Calhoun's behavioral sink that has so far been recorded in a population of wild primates. A legitimate criticism might be made of Calhoun's experiment, in that he had used domesticated laboratory rats whose behavior might or might not be duplicated in a wild strain; Sugiyama's appalling experience with the wild Dharwar langurs negates the objection.

Here in this western forest he found the extreme population density recorded by the three observers. Jay's had been less than 20 per square mile; Ripley's in Ceylon about 150; his almost 300. Whether density was the only factor contributing to social breakdown, Sugiyama could not know, but disorder was quite nearly perfect. There were territories, but borders were obscure and ill-defended. When troops met, leaders fought unassisted. Neither were there the rigid rank orders of dominance so characteristic of Jay's widely separated groups. Perhaps as a consequence almost all troops had only one adult male, though there might be six or ten adult females. Sugiyama speculated that without a hierarchy regulating the relationships of males, quarrels were so disruptive that only one male usually remained. The expelled males formed their own groups in the forest.

As we have seen, there is nothing unusual in disenfranchised all-male bands. The patas monkey in Africa, with troops consisting of a male and his harem, resemble precisely Sugiyama's langurs in India. But there is a very great difference. The patas organization is species-specific, and evolution of patas society has provided that surplus males be psychologically castrated. No such evolution has prepared the langur for the pathological social arrangements in the Dharwar forest. The loose gangs of surplus males are no more psychologically castrated than equivalent gangs of men. Overcrowding in the langur has produced a social situation for which langur behavioral evolution presents no answer. The results have been disastrous.

When the sexual season approached its peak, an all-male gang, with no more inhibition than Calhoun's prober rats, would descend on a troop containing females, kill or drive off the leader and any sub-adult males, and fight among themselves for sexual sovereignty. Far from mourning their departed overlord, the females would respond to the action with sexual stimulation which brought on an immediate peak of copulation with the conqueror. Infants were neglected. And the episode reached its climax when the conqueror bit to death all young.

Here is violent behavior on a grisly scale unprecedented in primate observation, and it seems a direct consequence of overcrowding. Of Sugiyama's nine troops under direct observation, such scenes of riot overwhelmed four in a season. The slaughter of the innocents, it is true, provides a form of population control. Yet I doubt that this is more than a by-product of social breakdown, of the triumph of disorder over order, of aggression no longer subject to social channels spilling over without inhibition into monkey immorality.

Similar situations have been observed in captivity. In another book I have described Zuckerman's baboons in the London Zoo who tore females apart, and Carpenter's rhesus monkeys, being shipped from India to the West Indies, among whom mothers fought with their young. Yet was it crowding alone? The late K. R. L. Hall once described to me an outburst of violence in South Africa's Bloemfontein zoo. In a single enclosure seventeen chacma baboons had lived for long in perfect adjustment. Then two strangers, one male, one female, were introduced. For two days nothing happened, and one cannot say that population density had been markedly increased. Yet tensions built. What sparked the incident no one knows, but when the fighting was over, only two of the entire group remained alive, and these were so seriously injured that they were destroyed.

The story of the langurs is persuasive but incomplete, for a pair of studies of the African vervet monkey yields precisely the opposite conclusions concerning crowded populations. The first was made some years ago by Hall's student at Bristol University, J. S. Gartlan, on Lolui Island in Lake Victoria. All human inhabitants of the island were evacuated early in the century when sleeping sickness was found to be endemic. The vervets took over, and when Gartlan made his study there was a population of fifteen hundred. While the island has an extent of eleven square miles, still only its fringes have the forests that provide choice vervet food. And so this wooded margin has been divided up into territories under an acre in size, each with a group of fifteen or twenty members. In terms of overcrowding, Lolui's vervets might almost be in a zoo.

Yet they live in order and in peace. Gartlan has described to me the territorial borders as so fixed that they might be painted with whitewash. Almost never did intrusion take place. When on rare occasion a few individuals found themselves in their neighbors' yard, they were either chased off with dispatch or vanished as soon as discovered. Fights were unknown. Early primate students, dedicated for mysterious reasons to the territorial virginity of primates, seized on Phyllis Jay's study to show that "langurs are non-territorial," and likewise seized on Gartlan's study to show that vervets too proclaim primate innocence simply because on Lolui they do not fight. Without doubt there was peace.

Then, however, Thomas Struhsaker published in 1967 his penetrating series of studies of vervets in the Amboseli reserve. No such limitations of space prevailed in this Kenya immensity. Groups were of about the same size, but territories were from ten to a hundred times larger. Yet they intruded with purpose, defended in concert, and fought with relish. As aggressive were these vervets in the vast spaces below Kilimanjaro as were their fellows pacific on the crowded island acres. Enhanced aggressiveness, too, entered domestic arrangements as well as foreign policy. The vervet boasts a remarkably colorful genital arrangement: his penis is red, his scrotum blue, and a little white strip of fur runs from scrotum to anus. The dominant male, to discourage an ambitious subordinate, gives him what Struhsaker describes with all charm as "the red, white, and blue display." Yet Gartlan in two years never observed it among his crowded island citizens.

The human inhabitant of that ever more crowded island in space, the earth, struggling to work in his subway trains with their thrice-breathed air or along traffic-clogged streets where air has become an industrial product, returning home after dark with the prayer that his skull will survive intact the next darkened doorway, will do well to look anywhere for helpful insights concerning his problems. But at this developing stage of animal understanding he must beware of scientific simplifications. Overcrowding was, with little doubt, a factor in the violent behavior of certain Indian langur populations. Yet it contributed to peace and to animal treaty among vervets on an island in an African lake.

We must look further. We know that the phenomenon of space impinges deeply on our lives, far more deeply than we formerly understood. But let us forgo the quick simplicities, however authoritatively they may pose, and learn for ourselves whatever we can.

3

Glen McBride is a strong, red-bearded Australian lecturer in animal behavior at the University of Queensland whose work commands increasing international notice. Much of his study has been directed to the benefit of Australian stock raisers and poultry growers. But just as Schjelderup-Ebbe discovered the universal principle of social rank in a commonplace Norwegian chicken yard, McBride's imaginative inquiries into the spatial arrangements of quite ordinary barnyard citizens may well be leading us into some general principles concerning personal space.

The Australian, who evidently goes to cocktail parties now and then, was struck by the way people pose themselves in a crowded room. Almost never do two people in conversation face each other directly at close distance. (There is an exception, of course, if one is male and the other female and ideas are lurking.) But most men have had the uncomfortable experience of backing away from an aggressive male speaker who keeps pressing his face too close. We end up usually with our heels jammed against an implacable wall, and in our hearts a question as to why we ever came to the party. Yet if we sit side by side on a bench or couch, no matter how uncomfortably we may be jammed together, we speak freely and without self-consciousness. And if we are standing, then in a crowded room we shall make a satisfactory accommodation. We shall stand with our bodies at an approximate forty-five-degree angle to each other. And if one is forced to move by the passage behind him of a tray of hors d'oeuvres, then the other will unconsciously shift to retain the forty-five-degree angle.

McBride started taking pictures of hens in a fairly crowded chicken yard. His pictures were taken from above at intervals of seconds. And he found precisely the same arrangements where hens are pecking food from the same pan. There will be the similar angle. If one shifts, the other shifts. Only a highly dominant bird will move in, like the aggressive speaker at the party, directly to face another hen. But then the subordinate will quickly adjust her angle. And out of his observations McBride developed a concept of personal space.

Surrounding each individual is a portable territory with its deepest dimension in front. As the robin will threaten and fight any intruder on his fixed bit of exclusive space, so we resent intrusion into our personal space. The size of that space will vary, of course, with the species. There are kinds like the hippo school, indeed, demanding none at all, although our kind is not among them. Space will vary also with sex, confronting males demanding most, females less, male and female least or, happily, none. Size will vary with rank, alphas commanding and receiving most. It may vary with season, activity, even time of day. But a demand will always be there. That the demand is greatest in front—what McBride speaks of as the field of social force—received fascinating confirmation in a flock of turkeys which he condemned to an unendurably crowded pen. Whereever possible the turkey faced the wire fence, gaining the space outside as at least seemingly his own.

When I wrote *The Territorial Imperative*, my object was to establish the territorial principle as biologists have observed and defined it, relating to an exclusive area of defended space. I indulged in few speculations concerning the application of the principle to areas less tangible than real estate. Every such speculation, I judged, would weaken the rigor of biology's conclusions. Yet it was obvious that with the conceptual capacities of the human mind the imperative to defend a territory has been extended far beyond fence posts and locked doors. Jobs, departments in an organization, jurisdictions of labor unions, spheres of influence whether in politics or crime are as jealously guarded as a warbler's acre. When I first suggested the subject of my book to C. R. Carpenter, he was amused: Why bother with animals? Why not just visit Pennsylvania State University for a few weeks and keep an eye on the faculty? And one must add with lament that none so obeys the territorial imperative as a professor of sociology denying biology's intrusive suggestion that man is a territorial animal.

Any consideration of the problems of human space must begin and end nowhere if we deny the territorial propensities of man. If man is infinitely malleable, as so many would have us believe, then urban concentration should offer no dismay. We can adapt to anything, even to the crawling masses of insect life. It is a proposition that few would accept. The territorial principle has been evolution's most effective implement in the distribution of animal space. And if man is a being biologically equipped with territorial patterns, then at least we have a premise to work from. Urbanization is deterritorialization in the classic sense of denial of land. But perhaps there may be conceptual substitutes or symbolic channels that will preserve our biological sanity. We may be sure, however, that we must somehow preserve no trespassing signs.

McBride was dealing with that most subtle of territorial manifestations, personal space, and he showed how even this inviolable area offers accommodation in chicken yards or cocktail parties through avoidance of head-on clash. His observations of men were of course impressionistic. In the meantime, however, an American psychologist, Robert Sommer, had been working on much the same problem, without benefit of chickens, in hospitals and an old folks' home. He found the distinct preference at small square tables for people to sit corner-to-corner. "They wanted eye contact but not so direct that they could not escape." In a college library study hall, with its long tables, the first to arrive would take an end seat; the second, the farthest seat away at an angle. As tables filled up, various offensive displays were resorted to. Belongings were distributed widely, as a Thomson's gazelle distributes fecal matter to repel intruders. When belongings no longer served, elbows were distributed, or feet.

Ingenious man has developed ingenious methods to ensure his biological privacy. Sommer's various studies have reminded me of the sophisticated techniques of a much-traveled California friend who on crowded planes has systematized the preservation of an empty seat beside him. Against the inexperienced air traveler one has only to place a brief case or coat on the seat. As crowding becomes more explicit, however, one faces the experienced traveler and such outworn tactics will not do. One may try body-juggling. If one stays on one's feet and through eternal rearrangement of belongings on the floor confronts the ambitious seat-taker with implacable buttocks, space may be protected. But if it becomes obvious that only one or two empty seats will remain on the plane, there is a single last recourse.

Huddle wherever you are, taking as little space as possible, while firmly and conspicuously grasping the air-sickness container before you. The more experienced your fellow traveler, the more surely will he avoid the adjoining seat.

Preservation of personal space need not be so shameless as the techniques of my California friend. Dark glasses will do. By shielding one's eyes behind a transparent but darkened wall, one creates a space corresponding to McBride's field of social

force. That the custom of wearing dark glasses even in the murkiest of night clubs was introduced by Hollywood stars suggests the determination of the alphas to command maximum space. That the custom has significance beyond contemporary fashion finds its testimony in Robert Murphy's study of the

Taureg people, *Social Distance and the Veil.* The Tauregs are those nomadic "blue people" of the Sahara whom the traveler may, if he is fortunate, encounter in the market of Marrakech. Skeptical indeed one must be of any conclusion that the Tauregs, drifting occasionally out of their normal ranges beyond the Atlas Mountains, have in their ways been inspired by Hollywood invention. Yet the veil before the face protects the personal distance of the alpha Taureg precisely as dark glasses protect the space of the western alpha. I shall watch you: you will not watch me.

We are only beginning the investigation of biological privacy. It exists. Yet in our extremities of overcrowding we know little about it. Irwin Altman, like Sommer, is one of our rare psychologists distinguished by their habit of investigating what matters. The United States Navy, with an understandable interest in what matters to men in confined spaces, has been his patron. And Altman and his associates with sadistic glee have devised a series of space-torture chambers in which to confine shanghaied naval "volunteers." The space has been normally a room twelve by twelve feet in which the sad sacks might be confined together for periods of even three weeks. In such an area territories were quickly established: this bunk, that chair, this side of the table. The more quickly could such commands of personal space be established, the greater was the chance of success in a relationship. But matched pairs of sailors were studied emotionally before commitment to experiments. And a hypothesis of fair predictability was established: Two men, each highly dominant, could not last for more than a very few days. In a confined space the marriage of men, not unlike the marriage of man and woman, rests on the willingness of one or the other to accept a subordinate role.

Altman's experiments are still in progress, and their broader conclusions rest with the future. So do the observations of a clinical psychiatrist, Aristide H. Esser, today working in the mental hospitals of New York State. Like the wild men of African anthropology, Esser is a wild man of contemporary psychiatry, and he looks the part. Half-Javanese, Dutch-educated, he approaches the inhibitions of western psychology as an Attila once approached the bastions of Rome. As an adventurer he could not be more admired by the layman; as a professional, strangely enough, he could not be more admired by his colleagues. Esser is pioneering the investigation of personal space in relation to the mentally ill or deficient.

Almost half a century ago Eugène Marais projected the thesis that with insanity, and reduction of control of one's actions by the cerebral cortex, a reversion to uninhibited animal behavior takes place. Esser has found that in mental wards patients quickly form hierarchies little different from primate ranks. At the top, patients are quite free in their movements, confident, capable of receiving positions of trust. Lower, however, the disposition is to retreat into a small fixed territory where the patient will be found at least three quarters of the time. A nurse, approaching such a patient simply to light his cigarette, received a blow in the stomach. She had violated his personal space. By careful observation of individual patients in terms of dominance and territory, the arrangements of the mentally ill can be adjusted for the maximum comfort of all.

Even more revealing has been Esser's work with children so mentally deficient that they have been institutionalized. Among such children the problem of aggression has been a subject of much debate. Do reprimand and punishment reduce or enhance aggression? Experience has been ambiguous, seeming to point both ways. Esser found an answer. Such children almost always establish fixed territories. Inside his territory, the reprimanded child will cut his aggressions by half. Even the most hopelessly deficient child can learn in relation to a location he regards as his.

What we are just beginning to learn about the human significance of personal space stems almost entirely from the observations of Heini Hediger, who first established the concept of individual distance. On a March day in 1938, in Zurich's Bellevue Square, he noticed black-headed gulls sitting on the lakeside parapet. The gulls had arranged themselves evenly, just two railings apart. His curiosity caught, Hediger began watching other species. Flamingoes demand twice the distance, about two feet, whereas swallows will settle for six inches. All these and many others are "distance species," but there are also "contact species" with quite opposite inclinations. Hippos I have described in another work. Tortoises, some monkeys and lemurs, particularly at rest, even owls and spiny hedgehogs will crowd together in an animal pile. But such species are a minority.

Hediger's individual distance has become among students of men what I have referred to as personal space, the portable territory. The master zoo-keeper extended his observations to animal training. The distance an animal demands between himself and an enemy is of a special sort called "flight distance." This distance too tends to be species-specific, small animals demanding less, large animals more. But flight distance will vary according to accustomed conditions. In a protected African reserve, baboons normally may not move away until you have come within twenty yards, whereas in unprotected areas they will take flight at a hundred. It is the reduction of flight distance that is known as taming. When it reaches zero, the animal is a pet.

In the training of a circus act—with lions, for example—it is not to the trainer's interest to tame his charges too thoroughly. Instead he reduces the flight distance to a critical, predictable space. By entering that space just a few inches, he drives the lion in retreat with whatever decorative whip-snapping or chair-wielding may appeal to the customers. But when the lion encounters an obstacle such as a barred wall, the lion will whirl, snarl, roar in most impressive fashion. And now, if the trainer remains within the critical space, flight will shift to attack. The trainer, we assume, knows his business and he will have available an obstacle, such as a pedestal, which the lion must surmount to attack him. And at that instant when the lion is on the pedestal the trainer will step back out of the critical space. The lion will sit back on the pedestal, having gained his point, and the trainer will favor the audience with a deep bow, having gained his.

How perfectly such concepts gained from the behavior of the animal translate into the behavior of men has been demonstrated by the distinguished Columbia University psychiatrist Augustus Kinzel, in a study of inmates of a Federal prison. His paper was presented to the annual meeting of the American Psychiatric Association in 1969. Kinzel chose among the inmates eight with histories of maximum violence and six who, however dark their sins, had least often indulged in violent behavior. What did personal space have to do with it? In a bare room floored by two-foot square tiles, offering quick estimate of distance, he placed his subject in the center and from various directions advanced on him, slow step by slow step. The prisoner was instructed to cry "Stop!" when he felt that Kinzel was coming insufferably close. By repeating such trials the experimenter mapped the personal space of his subjects. The nonviolent group exhibited with fair accord a demand for no more than ten or so square feet. And, like the hens, their zone was somewhat exaggerated in front.

It was with the prisoners boasting a life-time record of violence that the startling departure was made. They reacted with clenched fists to an invasion of space almost four times as great as the non-violent inmates. Crowdedness, in other words, brings forth from some individuals a reaction far more violent than from others. These were the men who at a McBride cocktail party, heels against the wall, would have smashed the aggressive conversationalist in the face. But there was far more to it than that. The personal space of the violent inmates was *skewed to the rear*. While one must wait for experiments with a larger number of subjects, Kinzel's observations are too clearcut to be dismissed. Prisoners with the longest records of manslaughter and assault sensed approach from the rear as a threat quite distinct from that sensed by the non-violent.

What did it mean? Was it paranoia? Was it a specific dread

of the pederast? It seems dubious. Neither the psychiatrist nor this observer can give an answer, but the violent are prepared for violence. With repeated experiments over several months, and with growing confidence in the investigator, the personal demand for space on the part of all inmates dropped by about half. The rear-projection demands of the violent types dropped even more rapidly. Yet a demand for space remained, whatever the confidence and familiarity.

Kinzel's experiment confirmed the reality of personal space in man, measured it as the trainer might measure the space of his lion, affirmed the innate resentment against the intruder, and exposed a spatial contrast between those who react violently to life situations and those whose demand for space differs. Why do some men kill while others do not? Human diversity, even in response to space, must enter our reckoning. As we approach the stranger on the dimmed street, we must feel a cold wonder. Will his personal distance be approximately like ours? Or will he be different? We cannot be sure.

4

So new is scientific concern with the spacing of animals that discussion has not yet evolved reliable terminology with definitions agreed on by all. Hediger's "individual distance" has become "personal space" and, as we have seen, may vary with individuals in its dimension. Although Murphy applied to his Tauregs the phrase "social distance," personal space has become more generally accepted. Likewise it is widely accepted that personal space may vary with seasons. The lapwing and chaffinch may gather in quite tight little flocks in winter, but with the first sexual stirrings of the breeding season disperse to scattered, relatively large, strenuously defended territories in which the demand for space is maximum. In our examination of the human crowd we have at least a language for individual con-

tacts. But for the contacts of groups, so pressing in our demand for human understanding, we have few such terms.

Besides individual distance Hediger gave us the term "social distance" to describe the farthest point to which an animal will go as it strays from its group. The social distance of a group, in other words, may be expressed as a measure of its scatter. A baboon troop may scatter widely as it feeds, exhibiting its maximum of social distance. But as dusk draws on, distance will rapidly shrink until all go to sleep in a few close trees. Social distance, then, is a measure of animal need for the familiar, and this too is a term we can agree on. But for the opposite and probably more powerful social force, the rejection of strangers, I can find no term in the literature.

I dislike and regard as an arrogance the contribution of terms to established sciences. But since this discussion cannot proceed without such terms, I shall refer to the social rejection of strangers as animal xenophobia. And even as personal space refers to the portable territory surrounding the individual, so I shall speak of social space as the similar area surrounding the society.

It is social space that must first concern us. If it is an area of permanent location, with fairly clear boundaries defended either by fighting or display, then it is a territory in the classic sense. Such is the social space of Ripley's Ceylon langurs or Struhsaker's Amboseli vervets. If it is an area of permanence whose borders, whether clearly or vaguely defined, need never be defended because they are never violated—as is true of Gartlan's vervets, or of the savanna chacma baboon—then ethology is in doubt as to whether or not it should be called a territory. So far as social space is concerned, the problem is one of semantics, since in either case the group has its fixed, exclusive property. But then comes a different sort, for the area, like personal space, may be mobile. India's populous rhesus troops shift their location from time to time, but frantic antagonism will be their response to any strange troop or individual who comes too near, violating their social space. Between two troops observed by Southwick, which had split from a single original group, antagonism was as great as between utterly strange bands. And finally, as we have seen in Jay's langurs of the central Indian forests, space may be so unlimited that antagonism is unnecessary and simple avoidance gives the group its integrity and psychological elbow room.

If we think, as I have suggested, in terms of social space, then many debates between specialists lose their heat. Is a species territorial or not? If territorial behavior comes about by innate command, then why, as in langurs or vervets, do we witness such disparity of display? The innate command is for preservation of social space. And the local population will draw from its genetically available bag of tricks that behavioral pattern which is suitable to ensure it.

Animal xenophobia is as widespread a trait among social species as any single trait we can study, and reasons for its incidence abound. There are the genetic reasons which I have explored earlier in this investigation. Were groups to mix freely, then any genetic advance would be nullified in the pool of vast numbers. Closer to our consideration is the problem of order and disorder. A limited group of familiars, many of them perhaps with kinship relation, know each other as individuals. Each has learned what to expect of his neighbor, and in such groups sufficient order comes easily. But the stranger presents a problem. Infrequently —most infrequently—his admission will be tolerated. If it occurs, however, normal procedure will probably condemn him to the bottom of social rank where his potentiality of social disruption will be minimum. Xenophobia guarantees the integrity of the group and the least possible chance of disruption.

Except for a few monographs by perceptive students of animal behavior, xenophobia has been little written about. And I shall join in that neglect until we debate the origins of human violence. What we consider here is the proposition that xenophobia, animal or human in its manifestations, keeps socially integrated groups separated. Whether that separation is accomplished through active territorial defense, through sophisticated acceptance of territorial rights, through active group antagonisms wherever space may lie, or as in a herd through indifference, group identity is effected and social space affirmed. It is an animal rule, and it has reasons. We may find it a human rule as well.

One would think that with all the studies that have been made of densely packed slums, little could be left for discovery. Yet two observers, Edward Hall, of Northwestern University's anthropology department, and the University of Chicago's sociologist Gerald Suttles have in the past few years found and traced invisible territorial boundaries. What in Chicago would seem to any eye the endless, amorphous, indivisible geometry of the South and West sides—an appalling geometry to one who, like myself, grew up in it—becomes through close inspection a mosaic of territories as discrete as those of the Ceylon langur. And they contribute to social order.

Hall's most recent field of study has been the black neighborhoods. Since he is white and our times are incendiary, most direct observation has been done by his Negro graduate students at Northwestern. And what they found were territories each normally including two adjacent blocks. Boundaries fell at the middle of the surrounding streets. You might have a friend around the corner in your territory, but not across the street. Fellow inhabitants knew each other, whether personally or through gossip. Any nearest adult had the right to punish a misbehaving child, whatever family he might come from. Anyone regarded as undesirable attempting to move into the territory was subtly discouraged; if he succeeded, then perhaps with less subtlety he would be convinced that he should leave. A definite although unofficial structure of authority invested the group. One or several individuals respected by all mediated quarrels, made final decisions concerning social attitudes. The rotting, packed Negro ghetto was in truth a series of independent villages.

It was a time when the city of Chicago, with belated conscience and all civic urgency, was replacing the infested slums with hygienic, high-rise apartments. And Hall's group found that the apartments, destroying the old territorial social structure, were in truth factories of disorder. No longer could adults on their broken stoops bring neighbors' children to heel. Now the young vanished downstairs into space to form their gangs beyond parental reach. The perceptions, the quick communication of gossiping neighbors, the reaches of unofficial authority were lost in the mathematical anonymity and isolation of tall concrete honeycombs. So drastic were Hall's conclusions that the city of Chicago considered abandoning the high-rise as an answer to the problem of the slum.

Suttles worked in a quite different area, publishing in 1968 his *Social Order of the Slum*, one of the richest books in sociology's slim list. His district is known as the Addams area, since it includes the famous settlement house founded by Jane Addams toward the end of the last century. Since Suttles was investigating changes in urban life, he chose his district purposely as one with a long history of study. But some of the phenomena he encountered have probably changed little in the time; they have simply gone overlooked.

The Addams area itself has distinct boundaries, compresses about thirty thousand people into its decaying tenements, and, unlike Hall's neighborhoods, is of mixed ethnic groups—Italian, Mexican, Puerto Rican, and Negro. But a ghetto has risen within a slum. Shortly after the last war a public housing development replaced several blocks with a thousand-odd apartments. It was not high-rise. Even so, it is today inhabited exclusively by Negroes, the district's omegas, and they are confined to it. "Project living is one of the most permanent and inflexible forms of absentee landlordship." Suttles emphasizes the drabness, the uniformity, the lack of means of family identity, the estrangement of residents through suspicion, the oppressive homogeneity, the inability of friends and relatives to settle near one another and create a "little moral world exempt from the insinuations of the larger community." He writes: "There is steady appeal to force where familiarity and exceptional signs of trustworthiness or power do not furnish a clear indication of one's future safety. Acts of violence without material gain reach their apex among the Negroes of the area." Yet the Negroes constitute but 14 percent of the population of the Addams area. Hall's conclusions stand fairly confirmed.

"In slum neighborhoods, territorial aggregation usually precedes any common social framework for assuring orderly relations," wrote Suttles. Residents must memorize each other's personal character, even as would baboons. "It is advantageous for Addams area residents not to extend their loyalties or become too dependent beyond a narrow territory surrounding their own homes." Startlingly, he points to slum morality as a consequence of territorial loyalties.

Whereas we tend to associate the promiscuity explosion with

high-density areas such as slums, we fail to recognize it as a middle- and upper-class phenomenon. Suttles lived for three years in the Addams area, in the mid-1960's, and so one cannot dismiss his observations as obsolete. Yet he describes the honor, the discretion, the secrecy with which boys' gangs, territorially organized, protect the girl who has made an understandable sexual slip. Thereby they protect her prospects for marriage. He makes the astonishing estimate that out of the Addams area's thirty thousand crowded inhabitants, not over a dozen or so girls have made and accepted a reputation of being sexually available.

Another of his observations, quite acceptable to the observer of evolutionary behavior, may startle the reader conditioned to human uniqueness. I have described in another work the amity-enmity complex, a psychological phenomenon that unites and enhances the energies of joint defenders of a territory. The Addams area is a mosaic of small territories normally restricted to ethnic groups. Yet there is a larger territory, the Addams area itself. Parents warn their children not to cross, for example, Roosevelt Road. There is a general animal xenophobia exerting its force against those who live beyond the spatial boundary. And the consequence is a form of integration. Though the despised Negroes may be segregated in the public housing project, still there remains a conviction that Addams-area Negroes are somehow better than Negroes elsewhere. Much though antagonisms and intolerances may flourish among the various little ethnic enclaves, whether Italian or Mexican or Puerto Rican or Negro, still there exist certain working arrangements. If you are a thief, then you will probably practice your trade outside the Addams area. If you are a fighting gang of young Mexicans, then in all likelihood you will go looking for trouble with Mexicans from Eighteenth Street, not some local gang. "Moreover, there is the suspicion if not the certainty that all street corner groups within a territory might join forces irrespective of ethnicity. By their assumption that residential unity implies social collaboration, inner city residents may help create the situation they imagine."

The works of Edward T. Hall and Gerald D. Suttles represent for the concerned reader a Klondike of urban truths far removed from normal sociological dogma. As at the overcrowded cocktail party we make unconscious accommodations to preserve our personal space, so in the overcrowded urban district we make equally unconscious adjustments to preserve our social space. Even in the most humiliating of slum conditions, a certain dignity and order remain so long as territories remain possible. It is when social space can no longer be defined and guarded that terror appears at the door.

5

In the past few years a suspicion has been growing in the minds of some of our most thoughtful students of animal and human behavior that territory and social rank are aspects of a single innate force, the drive to dominate. In *The Territorial Imperative* I described them as opposite faces of a single coin. If the new thought is correct, then the description, while coming close, does not quite illuminate the relationship. Dominance over a piece of space—territory—would rather lie at one end of a long continuum grading into dominance over one's fellow beings. The absolute sovereignty of the robin over his acre stands at one extreme; the absolute despotism of the alpha hen in a rigid barnyard pecking order stands at the other. And in between appear many gradients of relationship.

Two of our foremost thinkers—both, it so happens, with leanings toward physiology—have independently explored the concept. One, an American, is David E. Davis of North Carolina State University. The other, a German, is Paul Leyhausen of the Max-Planck-Institut. Davis' many studies of the physiological consequences of overcrowding led him to propose that how dominance is demonstrated is largely a matter of space. The roomier the environment, the more likely it is that order will be achieved through territorial spacing; the denser the population, the more likely will it be that societies must turn to rank order as an organizing principle. Leyhausen, perhaps inspired by his own experience in a camp for prisoners of war, has been absorbed by the human and in particular the political implications of the concept. He sees dominance over a territory as guaranteeing the rights and the liberty of the individual, so that territorial behavior in man reinforces democracy. But with increasing density the role of absolute hierarchy increases proportionately. To a degree, hierarchy is always necessary as the essential instrument of law. But in the end "overcrowded conditions are a danger to true democracy which it is impossible to exaggerate."

We have inspected enough animal and human examples in this chapter to know that density alone does not determine the loss of territorial sovereignty. Adaptability—a kind of animal common sense if the term be permitted—may ameliorate a condition otherwise intolerable. The crowded vervet monkeys on Lolui Island no less than the crowded human beings in the Addams area have retained territorial integrity by renouncing the disposition to quarrel over it. Were monkeys or men under such conditions tempted to dispute their borders, there would be no energy left for such other delights as feeding and fornication. By honoring the neighbor's social space, freedom is combined with order. That such accommodations come about as unconsciously in men as in other animals is attested by Suttles' difficulty in getting any Addams inhabitant to explain the system to him. It was simply dismissed as "natural," or "the way things are."

We may find certain accommodations easing the pressure of overcrowding. This does not, however, weaken the concept of space as now being advanced. Nor does it subtract from Leyhausen's grim forecast that with sufficient human crowding we shall see the gradual emergence of an elite who through political and police power will come to control most social space and refuse to share it, leaving for the great majority living conditions comparable to those accepted by sheep. It is scarcely a forecast. In the Soviet Union the separation has been proceeding for decades, and the dachas of the political elite stand in contrast to the workingman's shared flat. Through the same period the American flight to the restricted suburb may be preceding that period when police protection must be invoked to ensure space's privilege.

Ingenuity, social traditions, cultural acceptances may in a variety of fashions shield us from pressure. In his *Hidden Dimension* Hall comments with humor on the Japanese capacity, through mental concentration, to shut out noise. For the western man, however, raised in a tradition of moderately thick, sound-buffering walls, a night in a Japanese hostel with its paper walls when a party is flourishing is an experience not easily forgotten. As a longtime resident of Rome, I can confirm Hall's observation. The Italian predilection for decibels is such as to press me to the conclusion that either a most improbable membrane protects the Roman eardrum, or something is wrong with me.

So we protect ourselves. But in the end neither Italian eardrums nor Japanese concentration can defend us from the tyranny of numbers. We may raise our children to enjoy and embrace the crowd; but, as Leyhausen has acidly commented, we can also raise them to enjoy alcohol or drugs. Whether the acceptance, the seeking, the abandonment to the mob is selectively adaptive or maladaptive in *Homo sapiens* is a question for which evolution will provide its irreversible answer. What we should consider now is the thought, today being explored within ethology, that all spatial arrangements reflect the innate drive of the individual or the group for power, in the most Adlerian sense.

Dominance has long been described as what happens when any two animals pursue the same goal; one succeeds in establishing the dominant relation, the other becomes the subordinate, and the relationship will determine without further quarreling all other rivalries. But in our most recent studies animal behavior comes to appear ever more complicated, ever more resembling our own. Just as a man may be dominant in his home, subordinate in his office, popular at the corner pub, and a rank-and-filer in the local political organization, so role plays a part in the life of social animals. The clearly alpha monkey may leave to another leadership of the group in its foraging for food, may lead or not lead a general territorial battle, tolerate without interference the sexual activities of a subordinate, leave to others the chasing off of strangers. Frequently it amounts to no more that the aloofness of the alpha that we have inspected in an earlier chapter. But frequently also there is a clear distinction of roles: the dominant animal in one social role may be otherwise in another.

Out of these studies of varieties of dominance one clear variable emerges: the relation of dominance to space. The starling is non-territorial, but it is dominant over all fellows in the vicinity of his nest hole. Fighting crickets defend no territory, but the winner will almost always be the cricket closest to his niche. Leyhausen has conducted intensive studies of cats, and found that the defeated tom, when he retreats to his home place, regains his confidence. One of the clearest of studies concerned rabbits in Australia. A warren is divided into group territories, each with five to eight adults and a clear linear rank order. The alpha may defend the territory so viciously as to kill an intruder. Yet if he enters another territory, he becomes a subordinate. The many observations have led the American William Etkin to comment that while a proprietor behaves like an alpha on his own territory, he behaves like an omega anywhere else.

Such have been the studies leading to the hypothesis that a territory is the consequence of dominance over a piece of space. The will to power is satisfied by real estate, and dominance over fellows beyond his borders does not concern the proprietor. If the hypothesis is correct, then we can understand why size of territory is of minor importance, and its significance to urban problems is considerable. You are as much an alpha on your own nest site in a crowded gullery as wildebeest bulls sixty to a hundred yards apart. And we may see also that in territorial systems there are as many alphas as there are properties, and all have what we might call political equality.

Returning our attention to man, we may sympathize with Leyhausen's forebodings concerning democracy. The rural peace which so easily inspires our nostalgias was not too long ago a spatial arrangement making possible a maximum number of human alphas. Whether farms were large or small, all proprietors were on an approximately equal psychological footing. But then as territories expanded, so their numbers shrank, compelling a reduction in the number of alphas. Pressed also by mechanization, the new class of rural omegas turned to the city. And while the city tolerated and encouraged territorial arrangements, they were of a quite different qualitative order.

The continuum of changing dominance from the robin's acre to the barnyard pecking order has been duplicated in the human experience. The urban concentration has come to present as its conventional prize not dominance over space but dominance over men. It is the status struggle—that rat race which rats under natural conditions do their best to avoid. Territory lingers, but as a symbol of status and not as a prize in itself. For comfort's sake we seek a large apartment, but were comfort all, then we should not worry about the fashionable address. It is a symbol of status and, while presenting a degree of territorial security, commands through size and location its far more pressing symbolic contribution to rank. Similarly, the size of a man's office may offer no functional value whatsoever except as an advertisement of status. Dominance relates little to space, but almost exclusively to dominance over our fellows.

The ambiguity of the automobile in urban life bears fascinated if morbid inspection. It is a mobile territory without doubt, confirmed by the readiness to defend it emotionally on the part of both the driver and the dog beside him. The territorial boundaries of bumpers and fenders likewise separate proprietors so that even in a standstill traffic jam one confronts a territorial mosaic in which proprietors, aside from fleeting rage, acknowledge not at all each other's existence. Perhaps the car is a tender subconscious keepsake from a time when we walked beneath trees that were our own. Yet in urban life the car is a monstrosity. While it provides for the proprietor a tiny area of spatial privacy and a carpet on which to fly away, still in the city it exposes him to human density at its worst. As a territorial prize worth gaining, the car could not exist. Only its value as a symbol of status converts the urban automobile from prison to prize.

Throughout two volumes of these investigations I have stressed that territorial behavior is an evolutionary mechanism of *defense*. I stress it again. It is aggressive in that the proprietor, challenged, will fight. But almost never do we find in the territorial principle the concept of conquest as we human beings know it. I have resisted, despite many a temptation, the extension of the territorial concept into those human symbols, apparently equivalent, such as money and position. I may be thankful today that I have so resisted, since the Davis-Leyhausen concept of dominance reveals the distinction between possession that defends individual integrity and possession that encroaches on the integrity of others. Space is the essential criterion. Given space, territorial arrangements make possible the invulnerable individual. Stripped of space that is his own, de-territorialized man is stripped of his invulnerability. And that, down below all the asphalt and concrete, is the final statement of the urban problem.

As in my investigation of territory I resisted the temptation to extend the concept beyond my own territorial boundaries of inquiry, so in this investigation I must resist any pressure to stray far beyond the area of the social contract. Tempting it is to become absorbed with the technological horrors of the city's material environment. The pollution of air, the noise in chorus vomited by too many men and machines, the foraging for lunch when everyone else is foraging too, the lunatic adventures of telephone systems, the queuing in London or the scramble in Rome, the indignities of transport, what happens when something breaks down: there is one side of the urban problem, the side of which we are all most aware, which comes about when the machine that has been a willing slave in our mastery of nature turns its jaws about and masters us. For problems of technology this inquiry offers neither premise nor promise. But there is another side to the problem of the city.

As gradually de-territorialized man loses the exclusive space that once protected him, so gradually he proceeds through continuum's tunnel into the urban battlefield. We lose space as an ally. We enter the arena where man faces man and alliances are temporal.

Few strains on the social contract can today compare with the urban challenge. Space has been devoured and can be recaptured only with the romanticism which regards the summer cottage, occupied four weeks a year, as territorial reincarnation. The urban environment demands that we compose our bodies, our movements, our fecal matter, our gaseous extrusions, our aggressions, our drunken excursions, our noisy adventures, our quarrels man and wife, parents and children, our sexual dalliances, our ambitions, our frustrations, our personal loves, our personal hatreds, our political affinities or political aversions into an urban whole which approximates social order. If I am correct in describing the biological social contract as a balance between necessary order and necessary disorder, then the urban challenge must be described, with most delicate understatement, as a large order.

But as I have projected the philosophy of the possible, one must pause. To meet the material problem of the city, virtually anything is possible. And certainly within this inquiry's confines I shall assume that for technological problems, however baffling, technological solutions for affluent societies must exist. We have a talent. Urban men, driven year by year toward maddened desperation, can and must find the will that put men on the moon. We have already the talent. But for solving problems of our emotional environment, evidence supports no optimism at all.

Space—not that between Earth and Mars but between man and man—remains today as great a mystery as were fire and water to the ancients. With ethology's hypothesis that decreasing space means increasing tyranny I am in accord. And perhaps it is an extension of the hypothesis that further disturbs me, for a peculiarly human fate overhangs the urban outcome. We compete, in Wynne-Edwards' terms, for conventional prizes. But as density increases, the prizes grow fewer.

For every territory there is a proprietor, an alpha who has won his conventional prize. Not even, of course, in the rural society of orchard and garden, small-town grocery store or dry-goods shop do all men become proprietors. Still the alphas are many and the prizes remain real. But as urban concentration grows and competition shifts from dominance over a piece of space to dominance over our fellow man, not only do we encounter the hierarchy of despotism from which territorial man was largely protected but we encounter a vanishment of alphas. Prizes grow fewer. And as we press farther into the ever condensing mass we find ever enlarging human organizations, be they corporations or labor unions, political groupings or taxpayers' protest meetings. And as organizations become larger alphas become fewer.

I have mentioned much earlier the plight of the maturing young, still invested with the demand to prove themselves as individuals, facing the immensity of organizations in which the individual risks annihilation. But now we must inspect the phenomenon from quite another angle—the reduction of alphas. Whereas in a simpler and more spacious time a thousand shops provided society with a thousand alphas, now a single chain of department stores provides one. The conventional prize of the status struggle becomes as unreal as a kingship in Kansas.

We are turkeys standing against an Australian fence, necks outstretched, gazing longingly at that space outside which never again will be ours. Yet even as turkeys we are cheated, since no turkey ever endured a situation in which the goal of alpha bird became a figment of turkey imagination. You and I make accommodations, it is true. We invent side organizations wherein we may excel: the parent-teacher associations, the readers' clubs, the dedicated groups of butterfly collectors, the enthusiasts for antique motor cars, the impassioned devotees of political movements, the walkers who pound down the countryside, the travelers who bring back improbable photographs of improbable places to inspire temporary omegas into feats of yet more improbable tourist accomplishments. But for the born alpha such satisfactions are not enough.

I have mentioned that within any population normal diversity of individuals leaves between 3 and 3.5 percent in a group that must be regarded as mentally deficient. Measured in terms of IQ, this means 70 or less, but all forms of testing give approximately the same spread. And while one may debate just how much of the IQ score may be regarded as genetic and how much induced by environment, still there can be little argument that an IQ of 70 is so low that even an enriched environment is unlikely to improve it beyond 80. And this is an IQ still so low as to guarantee that the individual will remain within omega ranks. At the opposite end of the symmetrical curve one finds, of course, the 3 or 3.5 percent of gifted individuals with IQ so high —above 130—that even a deprived environment is unlikely to drop them from the role of potential alphas.

In my discussion of the alpha fish I suggested that intelligence is far from being a final determinant of high rank. But it is a most important one, since it is unlikely that an unintelligent gorilla male will ever become the silverback who rules the band, or a dull elephant cow ever become the leader of the file moving silently through the forest. And while to my knowledge no studies of the subject have ever been made, still on the demonstrable basis of the deficient and the gifted, I consider it a most conservative guess that 5 percent of any human population consists of certain omegas, while a high 5 percent consists of potential alphas. In the American population this would indicate that there are ten million people for whom there must be room at the bottom, and another ten million who must have some hope for the top.

If you are a potential alpha growing up in the United States of America, this means that when you mature you must join the ranks of ten million other potential alphas competing for a di minishing number of alpha roles. Your aggressive potentiality is uncontainable within a society of such diminishing prizes. Equally severe, however, is the problem of the condemned omega. When men turned earth with a spade, moved coal with a shovel, cut hay with a scythe, carried hods on their backs, the body could support human dignity though brain be inadequate. Such work in a highly organized, highly technological society is all but gone. As opportunity for the demanding alpha is withdrawn, so refuge for the condemned omega ceases to exist. The frustrations of urban society reveal in nakedness what our earlier rural societies modestly clothed: the innate inequality of men.

Injustices visited upon the gifted are scarcely distinguishable from injustices visited upon the ungifted. The pursuit of equality, that natural impossibility, condemns to mediocrity the gifted. The expectations of status beyond hopeful achievement condemn to frustration the giftless. Human density as achieved by urban concentration provides, as in any system of natural selection, an audience for naked display concerning human worth. But as we know it today the city, failing to recognize the inborn inequality of men, provides neither sufficient prizes nor booby prizes for citizens condemned to the rigors of density's struggle.

In terms of the social contract it is the urban concentration, juxtaposing unequals within a crowded environment and without the protection of space, that places incontestable demand for other than romantic concepts of men. It is the urban concentration that commands reassessment of social views. The city perhaps in *Homo sapiens'* evolutionary course is like that hurdle in a steeplechase making impossible survival if this hedge, that ditch, cannot be leaped. Our philosophy of the impossible makes likewise impossible all but a fall into the urban ditch.

I do not know what our outcome may be. Nowhere in the history of animals can one find precedent for the urban dilemma. Nowhere in the history of man can one find instruction for the future of a Tokyo, a New York. You and I, like flying molecules within a microscopic area of space, must seek adjustments, alliances, dispensations, tolerances, mutual goals, mutual defenses, new rules of the game subscribed to by all, new opportunities of achievement underwritten by common consent, new sanctions for the violator, new compensations for the violated, new dreams, new dreads, new uncontested heroes, new uncontested dross. The city forces upon all of us whatever reason man can mobilize. So we go forward. Or so, like the pterodactyl, we vanish.

6

Civilization began in the city; and in the city it well may end. So impenetrable seem its problems in the later decades of the twentieth century, so predictable seemingly is the city's devouring might, so irreversible seem the forces that have pressed urban man into deepening dejection, that the throne tempts despair as its sole occupant. Yet I find an unreality in such pessimism since it denies our animal will to survive. If we are to despair of the city, then we must despair of the species that created it. And such a rejection I find premature.

The city is a cultural invention enforcing on the citizen knowledge of his own nature. Our fears must suggest that we are aggressive beings easily given to violence. Our intelligence, if we possess any at all, must inform us that we get along together more because we must than because we want to, and that the brotherhood of man is about as far from reality today as it was two thousand years ago. The city, that unreasonable invention, with thunderous voice announces that reason's realm is small, and that if human foresight were what it is presumed to be we should not be in the mess that we are in. Through belly-to-belly juxtaposition of beings, evidence is presented that we never have been and never shall be created equal, and that if man is perfectible he exhibits the utmost shyness in demonstration of symptoms. These—all—are considerations from which space tends to protect us. One must sympathize with Rousseau's rejection of Paris, and his yearning for those solitary woodland walks where the dream suffers least contradiction.

Such cannot be so in the city, for it is a hall of mirrors. Wherever we turn we must see a face, and the portrait does not please us. We flee. But in the city, spaces are few and mirrors are many. We proclaim the reflection false—it must be of somebody else. What perceptions we have we blindfold with obsolete doctrine; what brains we have we abuse with comforting lullabies that someone once sang us. Yet the city persists, for it is larger than we are. And with giant arms it presses our faces, as in a childhood nightmare, closer and yet closer to the looking glass. And someday we must say: This is I.

In the cities of the Euphrates and the Nile and the Maya lowlands, in the cities of the Greek islands and peninsulas and of the Italian hills, a civilized way of life came about. And we cannot know but that the urban pressures of the twentieth century may include as another irreversible course a movement toward human understanding, a quality without which we can no longer survive. Falsehood, by and large, is a luxury. The romantic fallacy concerning man's nature, lively and pleasing a companion though it may be on a woodland walk, is a thug at night on a city street. The richer we grow, paradoxically, the less can we afford such a mistress.

The problems of the physical urban environment should not, as I have said, lie beyond human solution. Neither should uncontrollably increasing numbers place beyond our powers an urban answer. As we saw in the last chapter, population numbers do not increase indefinitely. And in any event, the urban dilemma reflects not numbers themselves but concentration of numbers. Japan has achieved population control of perfect order, yet the delirium of Tokyo remains. In the United States one has only to fly from coast to coast across its immensity to know that our urban problem has come about through human choice.

The city is not a concentration camp. All but the most unfortunate of us become part of its numbers not as slaves, not as prisoners, but as volunteers. Just as Calhoun's rats freely chose to eat in the middle pens and thus created a behavioral sink, we freely enter the city and, faced by conditions of intolerable overcrowding, we are as free to leave. Our departure may entail material cost; but that too rests on human choice. Overcrowding is a voluntary condition to which each of us makes his contribution, finds accommodation, or freely rejects.

It is not overcrowding but the breakdown of social structure that calls forth urban disaster. No example is more illuminating than the violence and the regime of fear that invest American urban life. That regime is as perfect in a relatively uncrowded Kansas City as in the massed densities of a Chicago or New York. Were overcrowding the cause, then this could not be so. What we witness is the shuddering, drunken tottering of a social structure supported by an inadequate philosophy. We lie to ourselves. It has been our way. Our lies we find pleasing.

Space leaves room for liars; the city does not. Here we live because we choose, live with each other because we must. Here the urban magnifying-glass of eternal encounter reveals the next man as he is, and the hall of mirrors reveals us as we are. Here we make treaties in equity with our fellow men founded on human reality, and we preserve and perfect a workable social contract; or we fail, lose all, and accept the dispensations of the despot.

The city as we know it now and shall know it for decades to come is a giant test created by civilized man which his civilization must pass or fail. There is no reason to believe that we shall pass it. Or, for that matter, that we shall fail.

8. The Violent Way

Britain's *Guardian*—formerly the *Manchester Guardian*—is not only the best-written newspaper in the English-speaking world but is generally accepted as our most steadfast spokesman for the liberal conscience. That conscience was given new voice in the summer of 1969 when, like a verdant volcano, Northern Ireland erupted in a flame of religious hatreds. A pungent odor of lunacy pervaded the smoky scene, for issues seemed irrelevant, even old-lace. The *Guardian* pondered the Laocoön intricacies of grievances involved, accepted all, but concluded that Ulster was showing itself to be "a profoundly neurotic society," more the problem of the psychologist than the soldier or politician. A leading editorial titled "Man an Aggressive Animal" referred to the "paranoid minorities" that appear so commonly in conflicts throughout all the world:

> In recent years the study of human aggression, and of aggression in society, has attracted growing interest. . . . To know how the trigger mechanisms work in human society will not by itself avert explosions. But at least there is a better chance of avoiding conflict if we know what brings conflict about. In fact we have all been too casual about the hazards of our innate aggressiveness. It is something the United Nations could very well take up as a major international research project.

Neither the population explosion nor the density of urban populations, neither nuclear catastrophe nor the devious adventures of youth, represents a threat to our civilized future quite so perplexing as man's propensity for the violent way. Yet no aspect of human behavior is so confused by misunderstandings or so subject in discussion to bitterness and prejudice. The courage of the *Guardian* was to reassert the free mind's venerated liberal tradition of open investigation, and to confess that we truly do not know what violence is. Just what are the ingredients of human belligerence? We may have our hunches, but we must inspect them. We may hold to dogmas, but either we can reinforce them with evidence or we must deny them. If we are to have any success at penetrating our history's terror-strewn trails, then our one clear area of agreement must be that even as we cannot know where these trails will lead us, we cannot even be sure where they started.

The task of penetration is compulsory. If the social contract represents a delicate balance between a degree of order that the individual must have to survive and a degree of disorder which society must have to ensure fulfillment of its diverse members, then a significant ascendancy of violence from any quarter tends radically to revise the contract. The balance then will and must be maintained by force. No triumph of disorder can be other than temporary. When order has been destroyed by one force, so will it be restored by another.

A few years ago our eminent journalist and historian of the making of presidents, Theodore H. White, published a play called *Caesar at the Rubicon*. While Rome sank deeply into anarchy, Julius Caesar by agreement stayed with his powerful army beyond the river. The play of course concerns his final reluctant decision to break the agreement, proceed to Rome, and assume power. And it closes with a simple comment worthy of being stamped on the coin of every democracy: "If men cannot agree on how to rule themselves, someone else must rule them."

White's statement hangs like an ancient sword over the heads of all free peoples. Violence—to paraphrase Wynne-Edwards—is the pursuit of conventional prizes by unconventional means. When social partners can no longer accept the same rules and regulations, then violence becomes the normal pathway of departure. And it is a paradox that the more successful the violation, the more certain will be its ultimate failure. Order must prevail if men themselves are not to perish. But in the course of such reconstruction of the social contract, many a man has seen freedom perish.

We shall be wise to inspect violence now while our social contract yet permits diversity of opinions. Or we shall wait too long, and our contract will be lost, and we, the violators and the violated, will in silent agreement bow to a higher, invulnerable force. Then order will be all.

2

Since Konrad Lorenz and I, though scarcely the sole proponents of the view that aggressiveness is a quality innate in all living beings, have been the principal targets of environmentalism's aggressive attacks, it seems appropriate that one of us define a few terms. Although I cannot speak for the great Austrian scientist, I regard it as unlikely that he has concluded that war, piracy, murder, mayhem, blackmail, burglary and the clobbering of strangers are either essential or commendable activities in the human species. Neither can I believe that all of our distinguished opponents have cynically made use of that proven trick, the twisting of meanings, to discredit a witness. Instead it would seem to me that an honest confusion takes place in many a mind between what is aggressive and what is violent.

Recourse to Webster's Dictionary I normally regard as that dullest of devices occurring only when debate has exhausted all more stimulating plays and the players themselves are exhausted. Yet perhaps with a sigh we should begin with the dictionary and get done with it. In its distinction of meanings, Webster testifies that "aggressive implies the disposition to dominate, sometimes by indifference to others' rights, but now, more often, by determined, forceful prosecution of one's ends." And if we turn to "violent" we shall find: "Moving, acting, or characterized by physical force, especially by extreme and sudden or by unjust or improper force." With a shade more ambiguity, Webster offers for the noun "aggression" the definition: "A first or unprovoked attack, or act of hostility; also the practice of attack or encroachment."

Aggression, then, places emphasis on primary, unprovoked impulse, but leaves the question of force open. What is aggressive is the disposition to dominate, to seek one's ends whether or not by forceful means; what is violent consists exclusively of those actions characterized by physical force. When Lorenz writes of aggression, he considers the innate (unprovoked) drive to dominate. When in this chapter I discuss the violent way, I confine my objective to those actions implemented by physical force. But just how far confusion can be pressed may be read in a quotation from Leonard Berkowitz, one of our foremost American psychologists:

> Since "spontaneous" animal aggression is a relatively rare occurrence in nature (and there is a possibility that even these infrequent cases may be accounted for by frustrations

or prior learning of the utility of hostile behavior) many ethologists and experimental biologists rule out the possibility of a self-stimulating aggressive system in animals. One important lesson to be derived from these studies is that there is no instinctive drive toward war within man.

Berkowitz within an admirably short span of words succeeds in misusing almost the whole of aggression's vocabulary, misleading the reader as to ethology's conclusions, and capping it with a reference to that special category of violence, war, which appears as a sequitur to virtually nothing. In his *Human Aggression* Anthony Storr comments on the passage that "such a point of view can only be sustained if a vast amount of evidence from ethological and anthropological studies is neglected." We may note that it cannot even be sustained by the dictionary.

In this inquiry I shall attempt to discriminate as clearly as possible among three categories of conflict: There is aggressiveness, arising from the competition of beings without which natural selection could not take place. There is violence, that form of aggressiveness which employs or effectively threatens the use of physical force. And there is war, that particular form of organized violence taking place between groups.

Not for money and not for space, neither for women nor a table in heaven do men seek to best one another. We obey a law that, for all we know, may be as ancient as life on this planet. We seek self-fulfillment. Within the limits and the directions of our individual genetic endowment we seek such a state of satisfaction as will inform us as to why we were born. We have no true choice. The force that presses on us is as large as all vital processes, and were it not so, then life would return to the swamp. If there is hope for men, it is because we are animals.

This is the aggressiveness that many would deny. It is the inborn force that stimulates the hickory tree, searching for the sun, to rise above its fellows. It is the inborn force that presses the rosebush to provide us with blossoms. It is the force, brooking no contradiction, directing the elephant calf to grow up, the baby starfish to grow out, the infant mamba to grow long. It is the implacable force which commands the normal human child to abandon its mother's protective shadow and to join the human adventure.

The aggressiveness that commands us all, hickory trees or human beings, must from the moment of bursting seed or bursting fetal sac direct us to overcome obstacles. The gasp for air, the grasp for the nipple, or if we are a newly dropped wildebeest calf, then the shaky following of our mother, represents for all of us the first commandment of independent life: that we come to terms with our environment. And so, as our bodies are born, our drive to dominate comes into overt being. But the obstacles need not be physical. Whether lions or lemmings, should we be of the sort who arrive in litters, then we shall find ourselves from birth in competition with our fellows. Natural selection will begin. And in moderate probability the least aggressive among our litter mates will be selected out. If we are of the sort that comes one by one, as do normally monkeys and people, then severe competition will arrive more slowly. Undivided maternal attention will protect us for a while. But competition and conflict will come, whether with our siblings, with our parents, or most certainly with our peers.

We seek the sun. We pursue the wind. We attain the mountaintop and there, dusted with stars, we say to ourselves, Now I know why I was born. We win a Grand Prix or a Little League ballgame. Or we achieve a transcendent vision of heaven and earth and God. We find a scarred desk high-piled with old books and, enraptured, we discover in the musty past our shining selves. All is aggression. We live for, search for, spy, covet, connive for that thing sometimes inexplicable to others but of utmost meaning to ourselves. Some portion of space, real or symbolic, small or large, glorious or inconspicuous, we besiege, we assault, we capture and make our own: and in that conquest —or even in that flashing glimpse of an unconquerable peak—we fulfill whatever it is we are. Rarely, however, do we take what others do not seek.

Competition may be denied as a common event in the vertebrate world by as elegant a zoologist as Harvard's E. O. Wilson; but he refers only to competition for food or scarce resources. As sensitive a naturalist as Sally Carrighar may write that "nothing could so prolong man's fighting behavior than a belief that aggression is in our genes." But she is simply falling into the vocabulary trap of confusing the violent and the aggressive. An ethologist for whom I have the greatest respect, Britain's John Hurrell Crook, does little better. Doubting its innateness, he writes that "aggressive behavior occurs normally as a response to particular aversive stimuli and ceases upon their removal." If he refers to violent behavior, then he has a point worth inspection; but if he means what he says, then he reduces the processes of evolution to the circumstantial.

Aggressiveness is the principal guarantor of survival. Although like any other genetically determined trait such as mental or physical potential, individual aggressiveness must in degree suffer random diversity, still normal incidence should be sufficient to ensure the survival of populations and species. We may even accept Crook's reference to "aversive stimuli." It is the innateness of the aggressive potential which guarantees that obstacles will be attacked, the young defended, new feeding grounds found when old lie waste, that orthodoxies give way to innovation when environment so demands, that when social traditions rot in obsolete alleys social change will come about. It is the heart of the Lorenzian principle that without aggression as an inborn force, survival would be impossible.

But it is likewise at the heart of the Lorenzian principle that survival dictates aggression's limits. Without traffic laws, aggression is a drunken driver in a lethal midnight. As no population could survive without sufficient numbers sufficiently aggressive, so no population could survive were competitions customarily carried to deadly decision. And so has evolved throughout the species that body of rules and regulations of infinite variety which, while encouraging the aggressive, discourages the violent.

The problem of man is not that we are aggressive but that we break the rules.

Any species must risk extinction when aggressiveness finds its fences in ruin and violence an ever available entertainment. But social species risk most. When beings become biologically dependent on the group and existence is impossible without the cooperation of one's fellows, then the violent solution of natural disagreement becomes a form of suicide as emphatic as the migration of lemmings. That the human being exhibits a propensity for violence beyond any other vertebrate species is a proposition that none with a reading of human history will dispute. But that propensity must be inspected with care.

I have been discussing thus far aggression in Lorenzian terms and its compulsory appearance in the genetic endowment of any species with prospect of survival. When competent authority fails to distinguish between the aggressive and the violent, then constructive debate becomes a wasteland of words. But having made the distinction, we court further failure if we do not distinguish between two sorts of violence, since one is subject to the command of evolutionary inhibition. The other is not.

Civil disorder—the use of physical force as the final arbiter in the disagreements of social partners—has been the subject of evolutionary disapproval so long as social groups have existed. An environmentalist as confirmed as Britain's cultural anthropologist Geoffrey Gorer has written: "All known societies make a distinction between murder—the killing of a member of one's own group—and the killing of outsiders." Gorer speaks of human societies. But unwittingly he reveals the evolutionary continuity of animal and human morals, for there is no animal society that does not make the same distinction.

An abused cliché is the proposition that only men and rats kill their own kind. We have witnessed in the Indian langur what happens when the social order becomes incapable of restraining primate aggressiveness. The young, for reasons quite unapparent, get systematically bitten to death. We have witnessed, in the questionable conditions of captivity, what happened in the Bloemfontein zoo when two strange baboons were introduced into a stable baboon society and all wound up dead. Southwick, in his great Calcutta experiment, saved the lives of strange rhesus monkeys from the antagonisms of a stable society only by withdrawing them.

An illustration presenting us with an answer to why, in a natural state, non-men and non-rats do not normally kill one another came to me through a friend in Kenya, David Hopcraft, who on his farm has for many years explored the possibilities of raising indigenous African ungulates as a meat crop superior to cattle. There is a problem of fencing, since most antelopes can jump over anything. Granted the aid of a benevolent foundation or two Hopcraft enclosed a ninety-acre field with a fence of which any concentration camp would be proud. He then began his experiments with the most innocuous little animal on the African savanna, the Thomson's gazelle.

Hopcraft today, as a result of his many experiments, is a wiser man concerning animal behavior than when in relative innocence he began. The Tommie, weighing only about forty-five pounds, is an animal so small that the new but normal technique of capture by means of a dart injecting a tranquilizer is dangerous. The dart, fired with considerable force, may enter the chest cavity. And so Hopcraft, to begin the stocking of his enclosure, captured half a dozen or so Tommies by netting. Introduced to their new home, a male killed two females within ten minutes. Shortly not one remained alive. He had taken his animals from two different herds.

It may be argued that an enclosure is still an enclosure, whether a ninety-acre field or a laboratory cage. Yet today one may watch the peaceful field from a tall observation tower in its center. Almost a hundred gazelles share it without problem. But Hopcraft did not repeat his error. Furthermore, he introduced his males slowly so that each could establish a territory the defense of which would absorb his animosities.

It is the effect of natural arrangements, not the inoffensiveness of natural dispositions, that minimizes violent behavior in a natural world. Territory is perhaps the supreme peacemaker. Tinbergen records that herring-gull chicks, straying outside the family territory, will certainly be pecked and frequently killed by territorial neighbors. Latent violence is there. Antarctic skuas prey on their neighbors' eggs and young, resulting in powerful territorial defense on the part of parents. The effective spatial distribution and separation of animals, whether individuals or groups, may be just one of the mechanisms reducing the opportunity for violence in the natural world. But that men and rats are the only creatures who will kill their own kind is a statement of dubious validity in any consideration of the violent way.

Another cliché accepted too often even by continental ethology is the proposition that the more dangerous the animal, the more harmlessly will he ritualize and contain his aggressions. Impressive observations point to the conclusion. Eibl-Eibesfeldt, for example, has shown that poisonous snakes in combat will never use their fangs. A study of captive wolves demonstrated that the defeated wolf has only to bare its belly to the victor to inhibit further attack. But one may doubt.

Those ancient animals, snakes, have had eons in which to perfect their ritualized relations. Yet I recall the stunned observation made in Florida by Arthur Loveridge, of Harvard's Museum of Comparative Zoology, regarding the behavior of two coral snakes, the most lethal of American species. A triumphant friend there captured a twenty-nine-inch male coral snake and placed him in a vivarium with a twenty-three-inch male which he already possessed. They made a beautiful if deadly pair. Taking Loveridge to view his new prize, the friend found his triumph replaced by mystification. Only one snake remained, the twenty-nine-inch. Reluctantly it disgorged the twenty-three-inch, which it had swallowed entire. The victim was in perfect condition, beyond being dead.

Of wolves in the wild we know far too little, and one cannot doubt that ritualization within the pack represents normal behavior. Students from Purdue University, however, have winter after winter observed from light aircraft a wolf pack on Lake Superior's Isle Royale. Constant observation made the observers familiar with every individual, including the leader over several seasons. Then one day he was spotted with a limp, and shortly thereafter he vanished. They landed, searched the area where the pack had been, found a bloody stretch of snow. A few bits of fur remained, and there were broken bushes testifying to a struggle. The pack had not only killed but eaten him. And times were not hard.

I carried my doubts to Africa when in the summer of 1968 I made a general survey of predatory communities. Hunting dogs will kill and eat any member of their pack disabled in combat, and this perhaps was the explanation for the Isle Royale wolves. But Hans Kruuk and George Schaller, our foremost authorities on major African predators, both dismissed as imaginary any proposition that the more dangerous the animal, the more will he ritualize his aggressions. Kruuk, in the Ngorongoro crater, had witnessed territorial wars between adjacent hyena clans involving large numbers of animals in which no least restraint separated the aggressive from the violent. Since hyenas eat anything, those killed in combat present a certain problem to the survivors. Hyenas do not enjoy hyena flavor. The problem was solved by allowing the carcass to lie in the sun for a few days. By then, presumably, it no longer tasted like hyena.

Schaller regarded violent outcome within a lion pride to be an unlikely resolution of debate not because of ritual, but simply because members are too familiar with each other's capacities to test them. The relation between members of different prides, however, is a quite different matter. Strict territoriality normally prevents contact. But should contact take place, then the weaker's only dependable ritual is how fast he can run. Schaller's most sobering experience throughout three years in the Serengeti concerned such a contact.

Not very far from Seronera, the Serengeti's lodge, are two famous lion prides much admired by visitors. Each has a territory of fifty or so square miles. One boasts—or boasted—two large males and nine mature lionesses, the other three males and seven lionesses. Both have cubs and juveniles in plenty, adding up to impressive lion societies. In Schaller's first year, however, there was a stretch of dubious territorial boundary. And in the disputed area a lioness from the two-male pride one day killed a zebra.

One of her lords came along and, following lion propriety, took over the kill while she retired to await his appetite's satiation. But then two males from the three-male pride appeared. She wisely fled. Unwisely, he fought. They killed him. Not content, they returned later and found the three cubs of another lioness. These they bit to death. One male ate a cub on the spot. The other carried his away, in Schaller's phrase, "like a trophy." The third dead cub was abandoned. Schaller waited. The mother finally returned, found it, ate it.

I followed the story of lion violence, that season, around all of Africa and found not a major game reserve without a valid record. While Schaller's observation might be regarded as one of intertribal war, a recent Nairobi incident was one of murder, since it involved members of the same pride. Here a male with a magnificent mane was for long the park's hero. Then, for reasons quite unknown, he killed a lioness. The park authorities were shocked. He was their principal tourist attraction. They decided against prosecution, and put him on what might be called probation. Before long he killed another lioness. And the decision was made to castrate him.

It was then that my friend Anthony Harthoorn, professor of physiology at the University of East Africa, was called in. As the developer of the process of tranquilizing animals through injection by a propelled dart, Harthoorn is famous throughout the world of animal conservation and his authority can scarcely be questioned. He supervised the operation. The great lion, sad to say, ceased to be a tourist attraction, for his mane promptly fell out. And sad to say, too, a year later he killed another lioness and so was destroyed.

Any inspection of human violence must recognize that in dangerous predators other than ourselves lethal propensities may appear. What motivated the Nairobi lion? I do not know. As bewildering a question was presented in South West Africa's Etosha Pan, perhaps the world's largest game reserve and certainly the least visited. Here lion violence could by most unlikely logic be attributed to overcrowding, for lions are relatively few.

Etosha itself is a giant pan, half filled with water in the average wet season, that resembles in size and shape the Lake of Geneva. Spreading around it is the park, adjoining the homeland of the Ovambo tribe just south of the Angola border. The park is precisely the same size as Switzerland itself. Though no Alps break its monotony of endless horizons, Etosha is a tourist's collector's piece. At countless waterholes one may watch rare species difficult to find elsewhere—the gemsbok, the dik-dik, the greater kudu in number. Even the lodge where visitors first normally arrive is a collector's piece, for it is an old fort straight out of *Beau Geste*, a relic of the times of German occupation.

An ecological setting so immense and so sparse cannot support the massive herds of prey animals that one finds in a Serengeti. And so lions are less numerous and their prides are smaller. Nevertheless, for several years Etosha had its attraction to reward any visitor's hopes. It was a pride dominated by two enormous males, and it could almost always be spotted, since they kept almost always together. Inevitably, I suppose, they became known as Castor and Pollux, and the stars they were of this African stage. Who took the greater pleasure in them, wardens or guests, would be hard to say. When South West Africa's chief of conservation, Bernabe de la Bat—who described the incident to me—visited Etosha from Windhoek, the distant capital, he invariably paid his respects to the two great lions as he might to two great friends. Then one day Castor killed Pollux. Why? None of Etosha's staff has ever come on a clue. I am assured by Schaller that sexual jealousy could not cause conflict between males in a pride. And, in any case, I am assured by De la Bat that no lioness was in heat.

As mysterious as the violence of men is the violence of lions. Having completed my season's visit to the animals, I stopped in Pretoria for a night before returning to Europe. And there authorities of the Kruger Park informed me of a lion fight the previous day. It had involved solely lionesses and occurred just outside the gates of a park camp before dozens of awed witnesses. Eight or nine lionesses had taken part. One was dead, several were seriously mauled. What the fight had been about no one knew.

Since lionesses defend territory against lionesses, never against males, one might guess that the Kruger incident was territorial. But I have learned my lesson about guessing. In *African Genesis* I described a well-witnessed fight which had occurred three years earlier between two giant male gorillas on the high slopes of Mount Muhavuru in western Uganda. (The ape, we are assured, is never aggressive, let alone violent.) The fight lasted for twelve days until one of the monsters died. Not badly injured, he died apparently, like the Glasgow rat, of defeat. No sexual motive for the fight was possible. I ascribed the fight, dubiously, to territorial conflict since there seemed no other cause. But then later Schaller's studies of the mountain gorilla in the same chain of volcanoes revealed no tendency toward territorial defense. And so I do not know why the gorillas fought, as I cannot be sure why the Kruger's lionesses fought, and I have no inkling why Castor killed Pollux or why the Serengeti males killed the cubs or why the lion in Nairobi Park developed Jack the Ripper tendencies. All may be understandable to lions, but not, as a human being, to me.

It may be argued that such evidences of violent behavior in dangerous animals are isolated and of small significance. But the evidences have been gathered from a relatively small number of animals regularly observed. I should suggest that the incidence of violent solutions among observed lions compares excellently with the murder rate in our most lethal and highly publicized cities.

By no means does such a suggestion return us to the nineteenth century's "nature red in tooth and claw." But neither should we accept the nice-kitty fallacy that today is becoming all too fashionable. Dangerous animals are so described because they are dangerous. And if we are to inspect the propensity for violence in the most dangerous of all species, our own, then we cannot—through a reverse interpretation of human uniqueness —pretend that we alone in all nature sometimes fail to resist the temptations of the violent way. Such a procedure replaces the over-simplifications of the nineteenth century with the over-simplifications of the twentieth.

Having presented what evidence I have for deadly quarrels among species boasting reputations better than ours, I find it time to turn to those two quite different expressions of human violence, the struggles among groups of social partners and the struggles between organized societies. And since I regard war as the least of the threats to the human future, may I be permitted to turn to it first?

3

In the summer of 1945, shortly before our present generation of students was born, American warplanes deposited nuclear bombs on two Japanese cities. The shock that leveled the cities spread its emotional shatter around the world. But following the first horror and the first awful guilt, many an imaginative mind began to ponder. The destruction of life and property had been of little greater order than that of the Americans' earlier and terrible fire raid on Tokyo, or the inexcusable destruction of Dresden by combined Allied air forces. Yet neither had ended a war nor afflicted a world's conscience. Not the act but the idea of Hiroshima was what rocked us. No imaginable action of whatever horror could shake us that deeply, since human history from its early beginnings had habituated us to the worst. What shook us was the unimaginable. The idea of Hiroshima was too new to behold.

In the year following, as some imaginations came slowly to comprehend the idea itself, I listened to many a speculation that the terrifying instant in Japan had indeed marked the end of general warfare. The struggles of armies which from civilization's earliest moments had provided mankind with its principal stimulation could play no part in a future world. The destruction of a Dresden, the burning of a Tokyo, had brought to history no prize beyond misery. But the incineration of two Japanese cities, terrible though the deed, had provided an idea that must wake from his dreams any future conqueror: neither victor nor vanquished could survive a weapon of such order. The Japanese price had been worth the purchase.

Such speculations went shortly out of fashion. Perhaps that unreliable figure, world conscience, condemned meditations so coarse. Or perhaps the cold war with its confrontations of superpowers and the growing balance of terror frightened us out of what wits we had. Not until fifteen years after Hiroshima, when I was concluding the writing of *African Genesis,* did I again confront the problem. By then, however, a different question had emerged in my mind: "How can we get along without war?" And I wrote: "It is the only question pertaining to the future that bears the faintest reality in our times; for if we fail to get along without war, then the future will be as remarkably lacking in human problems as it will be remarkably lacking in men."

I considered the possibilities of nuclear cataclysm and dismissed them as either unlikely or of merely academic interest. Cataclysm is not warfare, since among the dead one finds no differences; whereas warfare and the triumph of arms has through all our history been the final arbiter of the arguments of peoples. I wrote:

No man can regard the way of war as good. It has simply been our way. No man can evaluate the eternal contest of weapons as anything but the sheerest waste and the sheerest folly. It has been simply our only means of final arbitration. Any man can suggest reasonable alternatives to the judgment of arms. But we are not creatures of reason except in our own eyes.

I concluded that in far highest probability the easy answer of cataclysm would not be ours, that existence without warfare must somehow become our way. I compared man to the gorilla who in the heart of the Pliocene drought lost his forests and the boughs that had been the focus of his existence and descended to those bamboo thickets to which he was so ill-adapted. Could man, without his wars and weapons, survive?

Deprived of the contest of weapons that was the only bough he knew, man must descend to the cane-brakes of a new mode of existence. There he must find new dreams, new dynamics, new experiences to absorb him, new means of resolving his issues and of protecting whatever he regards as good. And he will find them; or he will find himself lost.

Slowly his governments will lose their force and his societies their integration. Moral order, sheltered throughout all history by the judgment of arms, will fall away in rot and erosion. Insoluble quarrels will rend peoples once united by territorial purpose. Insoluble conflicts will split nations once allied by a common dream. Anarchy, ultimate enemy of social man, will spread its grey, cancerous tissues through the social corpus of our kind. Bandit nations will hold the human will a hostage, in perfect confidence that no superior force can protect the victim. Bandit gangs will have their way along the social thoroughfare, in perfect confidence that the declining order will find no means to protect itself. Every night we shall build our nostalgic family nest in tribute to ancestral memories. Every day we shall pursue through the fearful cane-brakes our unequal struggle with extinction. It is the hard way, ending with a whimper.

So I wrote in *African Genesis*. And a decade later I find no persuasive evidence that my view was incorrect. A decade of declining fear of general warfare and consequent cataclysm has offered evidence, I believe, for another hypothesis of predictive value: Human violence, once fulfilled on the battlefield, is today being fulfilled in the city's streets.

There is a paradox involved. Organized warfare, although an exercise exclusively human in the vertebrate world, received in truth reinforcement from natural law, whereas social violence, the human expression replacing it, breaks every rule of social species. Intolerable though the damage of warfare might be, still it united societies, strengthened social contracts, and gave outlet for animal xenophobia. And if we are to prepare ourselves for any profound understanding of sabotage, riot, political kidnappings and assassination, then we should inspect carefully the concept of the stranger.

I have referred to the universality of animal xenophobia. The stranger is driven out of a group's social space and is physically attacked if his attentions persist. The howling monkey roars, alerting his fellows in the clan; the spider monkey barks; the lion, without ceremony, attacks. However the animosity for strangers is expressed, whether through attack or avoidance, xenophobia is there, and it is as if throughout the animal world invisible curtains hang between the familiar and the strange. What constitutes a stranger? Observers in California once experimented with valley quail who live in coveys without territorial attachment. Coveys tolerate each other within the same feeding range, provided that strict social space is respected. Alien intrusion on that space will immediately be resisted. The observers found that if a bird is taken from a group and returned within a week he will be accepted as a familiar. But kept away for five weeks, on his return he will be greeted as an alien. He has been forgotten. Similarly, they found that an alien who lingers about the periphery of a covey's space for about the same length of time will come to be accepted as a familiar.

One would expect that primates with sharper perceptions than valley quail would come to know and accept a persistent recruit more quickly. But it is not so. In his early observations of the howling monkey Carpenter watched the efforts of a solitary male to join a group. He was of course greeted with roars and pursuit upon first sighting. The performance was repeated day after day. But how long would it last? Sometimes for days he would disappear, or Carpenter would lose track of him and believe he had given up. But then again he would appear, and once he bore a fresh vivid wound as a reward for his persistence. Then slowly antagonism lessened. When the study was suspended, four months after the original contact, the immigrant was being permitted to remain on the periphery of the group. He was becoming a familiar, although he had not yet attained the status of naturalized citizen.

In his recent study of vervet monkeys in Kenya's Amboseli, Struhsaker kept careful notes on a parallel situation. For reasons unknown, an adult male left one group under observation in an effort to join another. It was November when he made his first foray and in uproar was put to flight. By the end of December, however, his presence was being tolerated, though ignored. Then within another month he was helping to defend his new group's territory. His success came more quickly than Carpenter's howler, but still the process was a slow one.

Most observers agree that in a successful group all members must know one another as individuals. Xenophobia, as I suggested earlier, becomes a force assuring that social partners will be familiars. Even when immigration is permitted, it is only after a long process of familiarization. Predictability of behavior makes group life possible. And when the unpredictable occurs in a familiar, the response may be as violent as to a stranger.

In the course of his long studies of the herring gull Niko Tinbergen used a net to capture gulls for marking. Netting, however, presented problems, since his approach to a colony would raise the alarm call and set the birds to flight. And so he devised a trap. First he disarranged the eggs in a nest while the parents were away. Then he retreated about twenty yards to a blind from which with a string he could spring his net trap. A parent, returning, would immediately lean over the nest to straighten out the eggs, and the net would be sprung. But an odd event followed. The gull, struggling under the net, would immediately be attacked and pecked by swarming fellow gulls. He was behaving strangely.

In *The Herring Gull's World* Tinbergen regrets that concern for the survival of his netted victim prevented his ever making detailed study of the attackers. To him the social attack on the strangely behaving seemed of broad significance:

> In human society, "primitive" as well as "civilized," a similar instinctive reaction is very strongly developed. It is perhaps possible to distinguish three steps or gradations of rising intensity in the social-defense attitude of the crowd. The first is laughing at an individual who behaves in an abnormal way. This serves the function of forcing the individual back into normal, that is to say conventional behavior. The next and higher intensity reaction is withdrawal;

the individual has made himself "impossible" and his companions ignore him. This, viewed from the aspect of biological significance, is a still stronger stimulus to the abnormal person to behave normally. The highest intensity reaction is one of definite hostility, resulting in making the individual an outcast, and, in primitive societies, even of killing him. In my opinion it is of great importance for human sociology to recognize the instinctive basis of such reactions, and to study them comparatively in other social species.

We should recall that the herring gull, living and breeding in its noisy, crowded colonies, is a species in which restraints on aggressive behavior reach something near perfection. Through territory, through postures of submission, through such a spectacular displacement activity as madly pulling up grass when frustrated angers boil over, actual fighting is virtually eliminated. But in response to the strange, all rules are off. Aggressive behavior in an instant becomes violent.

The rejection of the strange, whether the strangely behaving or the actual stranger, combines with Hediger's social distance —the maximum distance that a social member will stray from his familiar fellows—to effect social integrity in animal groups. As we have seen elsewhere, rejection need not be accomplished by such forceful means as the defense of territory, or by the emotional assertions of antagonism such as demonstrated by the rhesus monkey. Xenophobia between groups may very well be expressed by simple avoidance, as do the langurs in central India's spacious forests, or as do those exotic creatures in Borneo's mangrove swamps, the great-nosed proboscis monkeys. Here space is not so great. In seven square miles along a river eight troops, each of about twenty individuals, were kept under observation. There was never conflict. On one occasion two troops slept in a single area, separating in the morning. But passive xenophobia ensured that there was no contact at all.

Even physical avoidance and the maintenance by one means

or another of exclusive social space may be unnecessary in some species to express xenophobia. Aloofness may do. Phyllis Jay's langurs sometimes met at waterholes. They ignored each other. Psychological space prevailed. Baboons do the same. Family parties of zebra may seem to combine in single, endless herds on an African savanna, but each keeps to its own. So do African buffalo. And although a single alpha male exerts a sexual monopoly within a sexually mixed group, and although groups are so closely associated as to seem one herd to the unpracticed eye, still almost never does a male buffalo leave his group in an attempt to join another. The powerful lure of the familiar combines with the uneasy fear of the strange.

That animal societies are closed, and kept separated by distrust and antagonism, has been a worry to all utopians devoted to an ultimate brotherhood of man. For those so worried, a paper that appeared in 1966 was like a tranquilizer prescribed by a most respected doctor. The paper was called "Open Groups in Hominid Evolution," and it was written by a British primatologist, Vernon Reynolds, who with his wife had made an excellent study of the chimpanzee in the Budongo forest of western Uganda. In his paper Reynolds ably summarized known evidences of human evolution back to the separation of the hominid from the ape lines certainly twenty million years ago. His description of contemporary human societies could scarcely be bettered:

> Modern man is territorial and aggressive, hostile to and intolerant of strangers, and lives within an authoritarian social structure in which self-assertiveness and competition for dominance characterizes the successful male.

Reynolds then presents an argument that we must recognize: that such societies are of recent origin and culturally determined. For evidence supporting his conclusion, he looks to the social organizations of the great apes, our nearest relatives. The chimpanzee exhibits, in Reynolds' opinion, neither the xenophobia which I have been describing nor the social ranks which we have earlier considered. It is a loose society, an open society, in which strangers are greeted amiably, even with excitement. Chimpanzee life comes closer than the life of any other primate to the arcadian existence which we once attributed to the human ancestor. But the chimpanzee is not our ancestor.

The thesis presented by Vernon Reynolds has been acclaimed widely if uncritically, since it shores up the tenet of cultural anthropology that, since human fault has been culturally determined, it may be culturally corrected. The thesis offends no follower of Jean-Jacques Rousseau—another distinct advantage. But gremlins haunt the machinery of its logic.

The gorilla is as closely related to man as the chimpanzee. His society is as authoritarian as any in the primate world. While exchanges between groups take place more frequently than in monkeys, still Schaller's long study of the mountain gorilla revealed small change in nuclear arrangements. Gorilla society cannot be described as open.

About the orang we know little. They seem virtual solitaries, avoiding everybody. The male even avoids his wife.

Our knowledge of the chimpanzee is today based on three studies, all excellent, by Jane van Lawick Goodall, by Adriaan Kortlandt, and by the Reynoldses themselves. None confirms the open society. That Goodall was able to identify two thirds of the chimps that came her way in the many years of her famous study suggests—as it has suggested to Washburn and others—that what she was observing was a single society divided into shifting sub-groups. All were familiars. Kortlandt's observations were made in a banana plantation verging on a Congo forest. His was an undoubtedly discrete society divided into a nursery group of mothers and young, and shifting sub-groups of males, sub-adults, and childless females. On one occasion he watched forty, all surely familiars, raiding the banana plantation at once. The Reynoldses themselves, in their earlier paper recording their direct observations in a forest environment resembling Kortlandt's, guessed at a home range of six to eight square miles with only all-male bands venturing farther. In each region of three they estimated a population of about seventy. It would accord perfectly with the observations of Goodall and Kortlandt.

What the evidences suggest is that social distance in the great ape is fairly large. One may wander far from one's familiars. But there exists at some point a borderline, turning them back to their familiars, and that separates whole bands one from another. Perhaps memory in the great ape, superior to that of the monkey, does not require continual reinforcement. But there are other considerations denigrating parallels between ape and man.

The ape is powerful. His dependence on society for protection is so minimal that Schaller could not believe that the last male gorillas had been killed, as reported, by leopards on

Mount Muhavuru. Yet the hominid, our evolutionary ancestor, weighed rarely over eighty pounds.

The ape's relations are amiable, it is true, and if not egalitarian in the gorilla, certainly approximating the relationship in the chimp. But Goodall observed that when meat was at stake—when a chimp, for example, had caught and killed a monkey or a bushbuck—the dominance of the killer was total. All others sat about with supplicating hands, and while he dispensed a favor or two, he munched his meat in the alpha role, ignoring his inferiors. But the chimp kills infrequently. Our ancestors killed for a living.

And we come to the final consideration of Reynolds' thesis. Terrestrial man and arboreal ape have been environmentally separated for a good twenty million years, in terms of the challenge of natural selection. The chimpanzee is the product of one road, we are the product of another. It is a long time, even in terms of evolution, and to equate the end-products is a wishful indulgence. But beyond all such criticism must remain a final roadblock in romanticism's way. Among almost two hundred species of the primate family, few evolutionary failures can be compared with the gorilla, the chimp, the orang. Limited in adaptability, they are confined to small provinces of the earth's terrain. Powerful though they may be in terms of muscular advantage, intelligent though they may be in terms of laboratory tests, the great apes all approach extinction. The Age of the Alibi tells us that this failure is due only to man's incursions. Yet here is the baboon thriving beyond human enmity such as the great ape has never known. His brain is smaller. But the integration of all baboon minds into a social mind has been an accomplishment that the ape has never approached. As one cannot equate the life of the baboon with the lives of the great apes, so even less can one equate the history of *Homo sapiens*, the most successful of primate species, with the history of the apes, our most astounding failures.

Superior survival has been in baboon terms, just as in ours, a condition of a superior social contract. Were Reynolds correct in his description of the unique open society of the ape, he would still be describing a character of evolutionary failure opposed to a character of evolutionary success. But one must not neglect his "Open Groups" paper, since, whatever future research may confirm or deny, it represents the most informed support for an argument quite opposite to my own. If I dismiss it, I do not ignore it.

General evidence, I believe, supports the conclusion that xenophobia is a factor in the life of all organized societies, and that certainly in the primate family the open society does not exist. The stranger is necessary, and antagonism directed against him has a biological basis beyond wishful denial. The hostility assures that the group will consist of familiars. It unites the group through the process which I describe as the amity-enmity complex. If an animal society is based on a territory, then joint defense of the territory against an intruding stranger not only enhances energy but enjoins mutual trust and sacrifice, just as a human group defending its homeland must seldom have problems with the social contract.

Can we wonder that warfare, satisfying such natural demands, has flourished throughout human history? Can we wonder that man, like other social animals, carries within him a dual code of behavior? As Gorer suggested, there must be few human groups that do not distinguish between the killing of an insider and the killing of an outsider. The one is murder, and for it we may be hanged; the other carries a variety of distinctions the best of which is a medal. Washburn has presented a similar estimate: "Whatever the origin of this behavior it has had profound effects on human evolution, and almost certainly every human society has regarded the killing of members of certain other human societies as desirable."

Neither Gorer nor Washburn was referring specifically to organized warfare, but rather to the social tolerance of violent behavior so long as it is directed outward. Warfare as we have known it is no more than a cultural institution, like the home or the market place, providing multiple satisfactions for a variety of biological demands characteristic of social species. I find it unlikely that any institution so efficient in its satisfaction of a natural demand could ever have been abolished except by the character of warfare itself. But that is what has been happening in our time. That war has become impractical and therefore unfashionable is evidence that as an institution it is not in itself a genetic expression.

What we have in our genetic endowment is the rejection of strangers and probably the propensity for violence. These have not been abolished. All that is vanishing from the human scene is the institution that once provided satisfaction for both without damage to social integrity. And so, subconsciously, we provide an answer to the question "How do we get along without war?" We transfer energies once directed outward to the inward expression known as social violence. But such an expression presents an intriguing problem, for now we must invent strangers.

4

The future of violence is immense beyond conception, the richest crop today ripening in human fields. The old-fashioned patriot may sigh for days when national honor, or sometimes national survival, compelled young men to go forth and somewhere beyond the horizon sink battleships, return or fail to return from bombing raids, perform daring landings on unpronounceable islands, ravage somebody else's countryside with massive mobile armor, slug it out bayonet to bayonet in fanatically defended piles of rubble which had once been cities, maim or preferably kill an absolute maximum of strange less-than-human-beings whom one had never met anyhow, and, above all, to bring a little excitement into this normally boring world. But the old-fashioned patriot is in a rut; he fails to use his imagination.

I have paid, I believe, sufficient tribute to war as an outlet for violent enthusiasms. It has been civilized man's most popu-

lar entertainment, but it was never perfect. War had two grave defects. For one thing, it was invisible, and, for another, it was undemocratic. War was invisible in that for the vast majority of the peoples involved the action was somewhere else. I cannot deny that during the Second World War the mass bombing of cities corrected this defect to a degree. The British citizen, for example, may recall such affronts as among the best days of his life. Even with such minor advances, however, war remained to its last days in great part invisible.

But war also throughout history has been undemocratic. In

earlier days it was the privilege of a small professional class. In later times, as warfare expanded in scope, such professionals were no longer enough and so selected civilians were extended the privilege of participation. Still, however, there was discrimination. An aging general sporting ivory-handled pistols might occasionally appear where the action was, but the privilege of getting one's brains blown out was extended, by and large, only to that elite class, the young males. It may be true that the violent enthusiasms of the young exceed that of any other of society's sub-groups. Yet the propensity for violence, whether or not it is of genetic origin, exists like a layer of buried molten magma underlying all human topography, seeking unceasingly some unimportant fissure to become the most magnificent of volcanoes. Individuals may vary widely in their taste for violence, as they vary widely in their taste for avocados. Yet to deny its incidence in all human groups—male, female, old, young, the immature—is the most flagrant of discriminatory attitudes. Warfare's tendency to remain invisible was bad enough; but the restriction of its pleasures to a single human class was an oversight of undemocratic, even unforgivable, proportions.

That leftover wars of recent years such as that in Vietnam are unpopular, unreal, purposeless and, above all, unwinnable need cause small regret; the new expressions of the violent way will be neither invisible nor undemocratic. Violence will take place on your doorstep. Old, young, male, female—whoever you are, the equalitarian dispensation of civil disorder will present you with equal opportunity to enjoy the excitement of violent dispositions. No longer need you experience such pleasures vicariously. No longer need you resent the monopoly of youth over violent outlets. You yourself may participate in those elegant confrontations of life and death which formal warfare normally denied you. You yourself, whoever you are, may enjoy that most delectable of human tantalizations, the possibility when you get up in the morning that you will not be alive at bedtime. For all—not just for the elite young—boredom will be extinguished.

In any consideration of such a human future, you and I must make our approaches methodically. And since the most effective displays of civil violence rest on the antagonisms of sub-groups, let us first give thought to their arrangements.

In any animal society, unless it is very small, the phenomenon of sub-groups is present. In an earlier chapter I gave lengthy attention to the peer group, the young-of-an-age, which as it approaches maturity must challenge or find other means of assimilation into the ranks of established males. I indicated my belief that in today's human society the peer group is replacing the family as a fundamental unit of social structure. We are returning to the typical arrangements of almost all primate societies, whether as common as the baboon or as uncommon as the colobus.

The family sub-group, centered on the parents, has had an honored place in traceable human history. Within the primate line, however, its incidence is rare, and even in earlier pro-simians such as lemur species one finds only occasional expression. The family, normally polygamous, seems a standard grouping only in more distantly related mammals such as the horse and the zebra, certain rodents, certain predators like the lion. Birds seem most faithful to the monogamous expression but they are far away from our evolutionary line. In the human emergence any sensible guess would be that the family has been a latecomer.

Far more powerful throughout the species has been the all-male band such as one finds in elephants and most antelopes, in most deer, and, as we have just seen, in the chimpanzee. In his *Men in Groups* Lionel Tiger has stressed its significance despite the disruptive influence of the family throughout human history and almost all human cultures. The all-male band is a natural sub-group not to be neglected as we explore civil violence.

Frequently overlooked is the grouping of females, sometimes of strong if subtle influence, as in the howling monkey or the patas. The nursery group, whether in the elk or the topi herd or the chimp, may be of no more significance than what is left when the males decide they want no part of domestic responsibility. Yet in large primate societies like the baboon and rhesus a most definite tendency appears for certain females to form preference groups among themselves, even though the whole society be sexually mixed. Concerted female action—if we are to read as seriously as we should such a book as Betty Friedan's *The Feminine Mystique*—may provide an unexpected source of violent behavior in the human future.

What is demonstrated by animal societies is the structural necessity for sub-groups. If a society is to fulfill its obligation to members, if it is to provide education for the young, protection from external enemies, orderly relations among members, and psychological satisfaction for certain innate needs, then, like any organization, it develops a degree of departmentalization. And integration of sub-groups is facilitated not only by such factors as xenophobia and social distance, but by perfect communication. Everyone understands what anybody else says.

Animal signals, whether vocal or visual, are largely innate, and so there can be no misunderstanding their meaning. Some birdsong is learned. But if a gelada monkey lifts its eyebrows, revealing a yellow stripe, there is a threat. No one can misunderstand. If a wolf or a rhesus monkey carries his tail high, all know: he is an alpha. If a samango monkey in his treetop home chirps like a bird, then all become alert, for there is possible danger. Animal signals may be limited and inflexible, but they are precise and incapable of misconstruction. The enormous potentialities of human language, counted by many as the foundation of human success, carry with them no such integrating command.

It is a truism that the diversity of human languages and dialects offers more barriers to common understanding than all the diversity of human gene pools that heredity can provide. We may praise communication, but in the human species noncommunication might seem to a baboon anthropologist a more striking feature of our way. By no more instant means do we identify the stranger than by his alien language, his alien dialect, or his alien accent. Within a human society the speaking of a single language serves, as in any animal society, to integrate sub-groups. Yet so subtle is a single human tongue that differences of inflection, of connotation, of construction, of even vocabulary serve as well to separate sub-groups as to integrate them.

The prime foundation for the future of civil violence must be the invention of strangers. And the prime ingredient for the invention of strangers must be non-communication between those who speak the same language.

That strangers must somehow be created in our midst is a need so drastic that without accomplishment civil violence might well fail. The dual code that acts upon us is not of human origin and lies probably beyond human veto. Charity for a social partner might in the end so undermine our merciless determinations that social chaos would prove just as impossible as social utopianism itself. To succeed we must have strangers among us. We must have those sub-groups so targeted by dependable xenophobia that sympathy becomes impossible and only contempt and contusion will do. But to create strangers we must have non-communication.

For the moment I pass over those fortunate societies like the Canadian or the Belgian, divided already by differing languages. Their future, at least for a simple start, is assured. The challenge to human ingenuity, as I see it, is to create non-communication between sub-groups of a single society speaking a single tongue. And we must bow to the never ceasing resourcefulness of that triumphant creature, man. Already, without knowing what is happening to him or in any sense knowing what he is doing, man is exploring the new way.

Students and faculty, for example, could never have accomplished in our universities even their tentative scuffles of violence and destruction if either group had not been richly provided with non-comprehension concerning the other. The universities present an excellent example indeed, since here presumably the same language is spoken with fluency, authority and precision. Yet on many a campus the handicap has been overcome so overwhelmingly that non-communication flourishes and the supply of strangers becomes as bountiful as the enticements of violence may demand.

The achievement of the universities admittedly has been nourished by the achievements of the floundering home. A few old-fashioned families may still cling to a degree of affection, love and toleration but, as we all know, understanding between parents and children has been for long on the way out. What goes on in the parent becomes as mysterious to the young as what the young are up to remains a mystery to the parent. The peer group claims its own, shares its secrets, compares its miseries, extends its compassions, enthrones its ignorances, amalgamates its hostilities. The parent is a stranger. He shares his guilts and his angers with his kind, yet has no true sub-group to turn to. It is a lonely sort of fate.

One of the few flaws that I can see in the crystalline future of civil disorder is the improbability of organized violence between parents and young. While mutual non-comprehension may reach the most perfect level, while the creation of strangers has already attained conspicuous success, while hostility flourishes and conditions of downright enmity sometimes prevail, still something is lacking. And I believe it is because the young, organized into a legion by the peer group, face in that stranger, the parent, too contemptible, too pathetic, too bewildered an object for worthy attack. Whether mother or father, they are fading relics of another time when the reign of the family prohibited strong ties with others. Friends they might have had, or brothers, sisters, cousins. But the biological role of the independent family effectively prevented the encouragement of a parent class. And so today, understanding nothing of their children and almost as little of each other, parents present a woebegone face. Old battered shipwrecks floating on new seas, they tempt few guns.

I may of course be lacking foresight. The formidable all-male band may find itself reinvigorated by family disintegration and present the young with something to get violent about. For the immediate future, however, the significant contribution of the parent is to generate animosities which, consolidated by the peer group, may be redirected at policemen, university faculties, or other more satisfying targets. It is a contribution to be respected, even by the young themselves.

I am inclined to agree with those who ascribe to our capacity for communication a significant foundation for human triumph and the clearest distinction between human and other species. But I believe that one can demonstrate that our capacity for non-communication is quite as remarkable, and quite as essential if through violence we are to approach that final achievement, social destruction. There is another ingredient, however, without which we should probably fail.

The intricacy of sub-groups in human societies lies of course beyond the grasp of other species. Religious differences can play no part in animal life, just as racial divisions are denied by geographical separation. More elaborate by far, however, are the sub-groupings that have come about through division of labor. They are infinite: longshoremen and cattle ranchers, truck drivers and schoolteachers, firemen and fishermen, doctors, soldiers, priests, carpenters, actors, postmen, criminals: there is no end. And in every such group, just as in the peer group or the all-male band, members have more in common with one another than with the community at large.

The spectacular divisions of human society, together with the developing arts of non-communication, lend spectacular hope to the future of social violence. We may throughout the 1960's have watched with awe the latent violence released by religious differences in an Ulster or an India. We may with even greater intensity have watched the hostilities of white and black in America, or with lesser attention the animosities of black and Indian in East Africa, the little-reported struggle of black and Arab in the southern Sudan, the overwhelming massacre of Chinese in Indonesia. Religious or racial hostilities come easily to man. Such non-communication is a quality virtually built into our relationships, and the stranger is identifiable from birth. I suspect that it will be the conflicts of quite different groups that will provide us with the furniture of future dismay.

Let us remind ourselves that aggressiveness, natural to all living beings, is the determined pursuit of one's interests. And aggressiveness becomes violence only with physical threat or assault. Let us also reflect that civil violence (I do not enter into such personal expressions as wife-beating or house-breaking) involves always a minority, a sub-group by definition. Either a minority like the black in America resorts to violence to gain its ends, or a majority like the Indonesians confronting their Chinese crushes a sub-group to prevent the gaining of such ends. A sub-group may live in peace with its fellows; but ideal indeed is the society in which conflicts of interest do not exist. And I do not believe that the final vulnerability of human society looks to such major sub-groups as are formed by race or religion. Rebellions will be crushed, or accommodations will be found. The insoluble conflicts rest on division of labor.

In America blacks and whites, parents and young, students and faculties have demonstrated the workability of non-communication and the creation of strangers. But as the baboon learns quickly what a troop is eating, so we learn quickly arts pioneered by friends or enemies. And I find it difficult to believe that among those countless sub-groups created by division of labor there are not some rapidly learning their lessons. While the strike is an action of aggressiveness on the part of a sub-group pursuing its interests, it is not, as such, an action of violence. Yet borderline cases become more and more frequent. When firemen or policemen or doctors strike, the action leaves the whole society as vulnerable to physical damage as if assault had been staged. When in Britain a few hundred factory workers stage a wildcat strike without union consent, throwing tens of thousands of their fellows out of work, the suffering of workers exceeds the suffering of management. Can one doubt that non-communication is making its inroads, and that even the fellow worker is becoming a stranger in the eyes of the small sub-group?

We have had countless opportunities in the past, when communication, if not perfect, still existed, to set up such institutions as the labor court for the adjustment of undoubted grievances without recourse to the strike. With the exception of a very few smaller societies, we failed. Now the large society finds itself more and more at the mercy of an infinite number of sub-groups without whose services it cannot survive. The days are passing when there still existed sufficient compassion on the part of the sub-group to restrain actions too damaging. These were the days, also, when the threat of external enemies united us, and the exercise of general warfare provided outlet for violent disposition. Now violence turns inward. Now the art of non-communication inhibits compassion for our social partner. And it is unreasonable to believe that social inhibition will long restrain the violent ripping of our social fabric.

The ultimate vulnerability of every society—the more advanced, the more vulnerable—is division of labor. It is a sense in which the under-developed country, not yet committed to a course of extreme specialization, grasps at least for the present a lively advantage. Though many may die in conflicts of tribe or religion, yet the more primitive society survives as a whole. It is the industrial society, irreversibly committed to human interdependence, that must react as a single organism to the dagger of civil violence.

The *Guardian*, in its admirable editorial, referred to Ulster as a neurotic society and to raging minorities elsewhere as paranoid. I am not sure that I agree. A touch of paranoia is inevitable if members of a sub-group are to test their strength against the weight of a majority, and to risk life and certainly comfort against odds. Any sense of injustice must be aggrandized to a point where risk becomes acceptable. Yet I find myself doubtful that such social scenes should be characterized as neurotic, since we neglect a prime fact that the youth with a paving-stone in his hand is enjoying himself.

We enjoy the violent. We hurry to an accident not to help, we run to a fire not to put it out, we crowd about a schoolyard fight not to stop it. For all the Negro's profound and unarguable grievances, there has not been a racial outbreak in America since the days of Watts in which a degree of carnival atmosphere has not prevailed. I myself may have no great taste for Molotov cocktails; it is because I am timid, not because I am good. Suttles, in his work with juvenile gangs in the Chicago slum, found that an expectation that gang members would join in an action could be analyzed: stealing might attract a fair number, but the prospect of a fight would enlist almost all. Few studies of violent crime or violent gangs show neurosis as significant motivation. Were we truly sick societies, then I suggest that the violent way might be more easily containable; it is because we are healthy that we are in trouble.

Action and destruction are fun. The concerned observer who will not grant it indulges in an hypocrisy that we cannot afford. And unlikely it is that an attitude regarding a taste for violent action as a human perversion will make any great contribution to the containment of our violent way. Similarly, the observer who seeks nothing but earnest motivation in riot and arson, who looks only to environmental deprivation, neglect, or injustice to explain it—who, in other words, seeks wholly in the actions of the majority the motivations of the minority—may flatter himself one day that he was violence's most dependable ally.

The Age of the Alibi, presenting greater sympathy for the violator than the violated, has with elegance prepared us for maximum damage as we face a future of maximum civil disorder. A philosophy which for decades has induced us to believe that human fault must rest always on somebody else's shoulders; that responsibility for behavior damaging to society must invariably be attributed to society itself; that human beings are born not only perfectible but identical, so that any unpleasant divergences must be the product of unpleasant environments; that the suggestion of individual responsibility on the part of the social member or the sub-group for which he bears accountability is retrogressive, reactionary, Calvinistic: such a philosophy has prepared in all splendor the righteous self-justifications of violent minorities, and has likewise prepared with delicate hand the guilts and the bewilderments of the violated.

As one views the vistas of violent behavior rising before contemporary man like the swelling foothills confronting the traveler as he approaches the Rocky Mountains, a conclusion comes with too great ease that we are finished. Men can live neither with one another nor without. Extinction is on the cards. But any such conclusion is superficial. The social contract is an arrangement of biological validity. Like the sexual impulse or human diversity, it acts, in its balancing of order and disorder, to preserve the species with a power far beyond human predilection. What is at stake in our times is not the survival of man, but the survival of man's most rewarding of all inventions, democracy. Again I recall White's comment, after Caesar had passed beyond the Rubicon, that when a people cannot agree on how to rule themselves, someone else will do it for them.

There is visible throughout all nature a bias in favor of order. We have no means to explain it, and perhaps never shall. But the bias is there. A prejudice governs the movement of stars within galaxies, galaxies in their relations with others. Order commands the orbits of planets about their sun, moons about their planet. Order invests all living processes, and while hindsight may sanctify it, still order is there. Evolution and natural selection are no more than names for those processions which any may observe in the history of species. That animal treaties are honored; that baboons do not commit suicide in wars of troop against troop; that kittiwakes successfully defend their cliff-hung properties and raise their young; that lions and elephants restrict their numbers so that a habitat will not be exhausted by too numerous offspring; that lemmings embark on suicidal marches when there are too many lemmings, and snowshoe hares drop dead; that when species can no longer meet the challenge of environment, they must quietly expire: all such transactions of animals furnish simple testimony to the prejudices of order in natural ways.

Our recent investigations of democracy have been experiments, no more. And if human temptation to do violence, whatever its origin, lies in our social arrangements beyond voluntary control, then we may be well assured that anarchy will not be the winner. Order will be imposed upon disorder. And we shall return to more primitive political dispensations in which the citizen submits to violent impositions beyond his power to challenge, and keeps the peace because he must.

5

If an evolutionary approach to our various human dilemmas advances not at all a capacity to solve or ameliorate them, then it becomes an exercise in systematic pessimism. But if the acceptance of man as an evolved animal leads to hypotheses of a predictive value superior to those that have formerly guided us, then we may be permitted to hope that some doors at least which previously blocked our way may at last stand open. And even though the evolutionist's hypotheses may shock us, still we cannot apply to them the pain-pleasure principle: that if they give us pain, we shun them; if they give us pleasure, we embrace them. Painfully though an hypothesis may penetrate our habituated acceptances, still one legitimate question may be asked, Is it of value or is it not? Does it reveal or conceal?

I can find nothing shocking in the hypothesis that within a democratic society any tolerance of violence, whatever its original justification, that leads to the proliferation of violence leads in all likelihood to the death of democracy. It is a proposition that scarcely demands evolutionary materials to sustain. Anyone, for example, who survived the strikes of 1969 in Italy experienced not only the apprehensions of anarchy but its unworkability as a mode of human life. Then when in early 1970 the keepers of the Rome zoo went on strike for a single day—in Italy the children's holiday that in the Protestant world is known as Epiphany—and again at Easter, the observer must wonder. Just what are the final limits of non-communication? Here is a people that above all others worships the *bambino*—whether your own or the next family's or the Son of Mary—yet among them are those who will strike against children.

The Italians will, of course, develop a measure of self-restraint for which history exhibits no precedent: they will surrender certain freedoms of action, a step for which the future shows small promise; or another Caesar must and will someday cross another Rubicon. The predictive value of the hypothesis lies in the certainty of the final step and the encouragement of more moderate steps to prevent it.

I have suggested that the hypothesis demands no evolutionary confirmation. It was available to Plato. It is available to any sane observer of contemporary events. Yet, oddly enough, it is the honest environmentalist who must be most deeply stricken by gloom. Since no improvement of environment has thus far been accompanied by other than increasing incidence of violence both criminal and civil, any future projection must be one of helplessness and total despair. Only if one approaches the hypothesis from the viewpoint of evolutionary psychology does one begin to see light.

To speak of evolutionary psychology is to christen the baby before it is born. A principal contribution of my own investigations, however, falls somewhere within its unspecified dimensions. In *The Territorial Imperative* I introduced the hypothesis of innate needs common to men and all higher animals, and twice in this present inquiry I have made use of it. I believe it to be of diagnostic value if we apply it to the problem of violence.

I repeat briefly: There are three innate needs which demand satisfaction. The first is identity, the opposite of anonymity, and it is highest. The second is stimulation, the opposite of boredom. The lowest is security, the opposite of anxiety. Our innate needs form a dynamic triad. Achievement of security and release from anxiety present us with boredom. It is the psychological process least appreciated by our social planners, producing the bored society described by Desmond Morris in *The Human Zoo*. The bored society could not be a reality, however, were we other than the anonymous society stripped in large part of our opportunity to search for identity. Egalitarianism, that most admired of human possibilities by the unsophisticated young, and the shrinking establishments of enlarging organizations, despised by the young, in equal measure contribute to the defeat of identity's satisfaction. And so the frustration of the search for identity just as much as the achievement of security forces the member of a contemporary society into the unendurable area of boredom. We are pressed as in a vise between the achievements of security and the denials of anonymity. No way presents itself but stimulation.

Pornography and riot are of a piece. They are means of stimulation from which varied appetites may seek stimulation. By no coincidence at all have waves of shock and sensation arrived simultaneously on the scene of a bored society. The exhibitions of demonstrators raising the blood pressure of their close-packed groups and the exhibitions by fashion of the female body hopefully raising the blood-pressure of the passing male—their appeal is the same. The private titillations of the *voyeur* with his photographs from Denmark and the private hallucinations of the sense-stretching trip with its weeds from Mexico —these too are the same. Sexual adventures among adolescents, casual adultery among their jaded elders, the display of nudity and intercourse on the screen or stage, the robbing of houses by affluent burglars, the sense-assaulting decibels of electronic music paralyzing the brains of its listeners, the baring of waitresses' breasts for the excitement of timid gentlemen who would collapse in terror if further were expected of them, the public display of illegitimacy in the obstetrical arrangements of the famous, the public display of dirty feet by oddly dressed young men who can find little else of stimulation to offer, the public display in a literary context of ancient, commonplace terms of the street by authors with larger vocabularies available—all provide stimulation for shocker and shocked alike in bored societies with nowhere else to go.

The silent, weed-grown palaces of ancient stimulation pro-

vide evidence for its final futility, yet violence is of the same piece. It is exciting. It carries stimulation to both violator and violated, whether through the joyful hatreds of the one or the fearful rages of the other. A riot in Chicago is worth all the circuses that old-time emperors could provide. And like any other form of sensual shock, violence to retain its stimulation must proceed ever to stronger or more novel levels of expression. That adaptable animal, the human being, habituates himself too easily to any present situation. The film-goer, this year excited by normal intercourse on the screen, must next year be provided with fellatio. The demonstrator, in one year exalted by his mass marches, must in another incite police to violent confrontation. The postal striker, content in one season to paralyze business and domestic communication, must move on to staging massed demands at street intersections which hopefully will reduce all traffic to chaos.

Stimulation's progress rests not alone on the shocker, but also on the shocked, on the ease with which we get used to things. The peaceful march no longer excites us, the simple nude no longer attracts notice, the empty mailbox becomes accepted with a resentful sigh. The progress of stimulation requires in a sense cooperation of all parties. But the outcome of that progress separates in drastic fashion the sensualist and the violent. It is the weakness of sensualist shock that at some point it must cease to stimulate. Violence owns no such limitation.

I believe that as we refer to the triad of innate needs we shall glimpse more clearly the difference in actions seemingly so much of a piece. We are pressed into stimulation by the flight from boredom which itself has been induced by both security and denial of identity. But novel sensory experiences satisfy only our need for stimulation. Violent experiences tend to satisfy not only stimulation but identity as well. The participant in some new sensuous adventure may obtain a fleeting sense of uniqueness, of being different; the young explorer of drugs may feel himself superior to more conservative fellows and, without doubt, to disapproving elders. But the acquisition of identity can last only so long as the stimulus remains unique. When drugs become commonplace, he is just another hooked young person. In his own fogged vision he may retain an identity; it will be apparent to none other.

But the violent are applauded. Whether the applause be the praise of their collaborators or the condemnation of their antagonists, they are recognized, identified, released from anonymity. Whereas the sensualist achieves identity only in a private world, the violent achieve it in a public world or the stimulation is not worth the seeking. Excitement and recognition become one. Even though that world may have its mysteries, its secrets hidden like the privacy of a Mafia or an underground political conspiracy, still the whole threat is recognized and feared by the larger society, and within the dark principality of violence—united, one may note, by the amity-enmity force—individuals are dreaded, admired, hated, loved, ranked. Identity is achieved.

I do not discuss the violent solutions of individual disagreement. Although two thirds of all murders coming to the attention of Scotland Yard involve members of the same family, still the butchering of a husband by a wife, while a domestic indelicacy, is neither something new under the sun nor an eloquent harbinger of social catastrophe. That most peaceful of people, the Kalahari Bushmen, who regard violent dispute between members of a band as the coarsest of manners, accept the beating up of a wife by a husband or of a husband by a wife (whichever is the larger) as a normal episode in existence not unlike the rising of the moon. Individual disarrangements may be a product of our human disposition for violence. Like suicide, they may very well in their incidence be a symptom of social disintegration, of declining inhibition against the violent way, a reflection, indeed, of despair. But they are consequences at best, certainly not causes, of the social phenomenon we are investigating.

The violent way that I discuss here is the creation, by means of physical threat or assault, of dark little worlds in the image of the society of which they are a part and against which they transgress. Such a world upholds its own values, defends its own territory, establishes its own rules, praises its own alphas, scorns or ignores its own omegas, punishes its own traitors. Whatever the nature of this little world of criminal or political conspirators, of juvenile gangsters, of grim old dedicated members of the establishment of power, of rebellious militants whatever their grievances or purposes, still much the same processes of the sub-group prevail: There is non-communication, eliminating social compassion. There is xenophobia, released by non-communication to identify the majority of our social partners as the enemy. There is an illusion of central position, justifying one's own purposes as right and everybody else's as wrong, and providing a proper degree of paranoia. Righteous ends, thus proved, absolve of guilt the most violent means. And within this little world of lunacy a new fellowship blooms, a new communication flourishes, anonymity vanishes, identity again becomes possible.

That openly or secretly, consciously or unconsciously, we applaud violence's success can be inspected only in terms of human unreason and humanity's original sin. Yet we do. And as if subconsciously aware of our weakness, we find the principality of violent demand becoming a striking invention, still another tribute to the ingenuity of man in fulfilling his innate, seldom-recognized animal needs. The sub-group has transferred the attractions of old-time warfare to a contemporary social scene where the nuclear veto is impossible. But impossible likewise is the invention. We know that Rousseau's vision of a primal world in which solitary man was good, amiable, without need or animosity for his fellow, never existed in man's dimmest beginnings or can exist in man's most remote consummations. Likewise we know that the Hobbesian view of a primal world in which every man warred against every man could never have existed as it can never exist. Neither Rousseau nor Hobbes in those pre-Darwinian days recognized that the social imperative has always been with us and must be with us till the last human spark goes out. Nor has the social imperative ever so compelled us as it compels us today.

The violent sub-group even as it asserts itself defeats itself. The student depends on the postman if he is to receive his check from home. The postman depends on the milkman if his infant children are to survive. The milkman depends on the dentist, unless he intends to renounce toothaches. The dentist depends on the telephone system if the milkman is to make an appointment. The telephone system depends on God knows whom, black or white, if bells are to ring. And the black man depends on the ring of the telephone if he is to have hope of organizing any but an amateur riot. So it goes.

The intricacy of sub-groups in a modern, developed society

presents us with our greatest vulnerability. But the intricate web of interdependence should present equal dismay to the violent little worlds. The animal cannot stand alone, and least of all animals, modern man. I should sink into folly, however, if I presumed that in a time of advancing non-communication a proposition of such elementary logic could exert influence or inhibition on the actions of an animal renowned less for reason than for ruckus. And so let us return to evolution's playing fields, to their rules, to their regulations.

Governments may fall, societies will persist. There is no alternative. A new Caesar will cross a Rubicon; Italy will remain.

Technological achievement, while eliminating the need for economic security from the lives of growing majorities, condemns likewise to anonymity the same growing majorities. The search for identity will not vanish.

The violent sub-group, whatever its just demands or righteous protestations, seeks as a primary satisfaction the innate need for identity. Violence proposes present success, ultimate failure. The warring sub-group, denying the social needs of all others, denies in the end its own.

The evolutionary course stands as virtually an imperative. We have come to translate security solely in terms of freedom from material want. But walking across a storm-swept city to work when the busmen are on strike is not security. Climbing twenty-two floors to an office because electrical workers are on strike and elevators are stilled is not security. Being unable to telephone the doctor when your child is ill—even the worry that one will be unable to—this is less than the secure life. Being unable to cash your paycheck—be you carpenter, oilfield rigger, or busman or telephone worker, for that matter—because the bank clerks are out offers small security if the rent is due. And wondering, when your husband is unaccountably late from work, what might have happened down the street—it is the uneasy way. Or listening tensely in a suburban bedroom—were those footsteps outside or inside? They must have been outside, but why are things so quiet? It does not make for restful sleep. Or at the office the numbing news that your competitor has failed through incapacity to keep up production, while you know that you are as bad off as he is, and you have 800 employees who may fall victim if fifty strike. At home watching in horror on the television screen a riot on 18th Street—the blood, the screams, the unloosed ferocity—did your son go? for he said he might. Why oh why is your husband so late? Why oh why have you had not a word for three long months from your daughter at college? And your husband, stuck for two long hours in a traffic jam caused by a hundred or so persons in a seated demonstration at an intersection, is pondering emotionally the values of the police state.

The portrait I draw is a gentle one, exceeding in no detail present commonplaces of human experience. I include no flamboyant though quite certain projections of violence's future, neither mass ravage nor mass reprisal. What I have emphasized is simply the replacement of materialism's old insecurities by violence's new anxieties. The accomplishments of anarchy return the entire social body to preoccupation with the lowest of the innate needs: security. Whereas yesterday we wondered where the next meal would come from, tomorrow we must wonder where the next blow will fall.

The disintegration of security may take either of two roads. There is the new road which I have been discussing, the way of fear, of anxiety for one's person, the superb terror of that new unknowable, just what outrage may happen next. But there is also the quaint possibility of the old road: that the erosions of anarchy will dismantle our vast technological establishment. The immense industrial and agricultural complex which has banished or can banish economic insecurity from our lives rests, as much as any other social institution, on the web of interdependence. Without a high degree of order—too high if we take the long view—it cannot function. And so the old road may take us back to those scenes we thought behind us, through old villages of bankruptcy and bank failure, lowering productivity, diminishing exports, pay cuts, mass unemployment. It is possible.

One cannot predict the manner of one's own execution, whether by the hangman's noose or the firing squad at dawn. We seem presently determined on the one or the other. Yet present determinations in a species so famous for its foresight, so inadequately equipped to display its credentials, excite small trust. How profound is our propensity for violence? We must look into it. To what degree is our capacity of foreseeing dependable? History will tell us, if the younger of those among us live to read it. Will some quite unpredictable surge of common sense overcome *Homo sapiens* in his extremity? While it is improbable, it is not impossible. Such could be our evolutionary way.

The evolutionist may treasure a certain optimism concerning man, because he is an animal. It is not an optimism to be taken too seriously. Yet a student of order and disorder may fairly predict that long, long before we achieve such calamities as I have outlined, decision will have been withdrawn from our command. Whether or not we have the vision to see him, still he is there beyond the broad dark river. He broods, he waits, just as he has always waited. Neither tall nor short, neither broad nor lean, shadowy in outline, without distinction of feature, he wears an odd sort of hat and an old, old sword at his side. And if we do not act in time, then he will.

9. The Lions of Gorongosa

Hunting is the master behavior pattern of the human species. It is the organizing activity which integrated the morphological, physiological, genetic and intellectual aspects of the human organisms and of the population who compose our single species. . . . Hunting played the dominant role in transforming a bipedal ape into a tool-using and tool-making man who communicated by means of speech and expressed a complex culture in the infinite number of ways now known to us.

These thoughts were recorded by William S. Laughlin, professor of anthropology at the University of Wisconsin, in 1968. The year before, California's S. L. Washburn had presented a similar view:

Human hunting is made possible by tools, but it is far more than a technique or even a variety of techniques. It is a way of life, and the success of the hunting adaptation (in its total social, technical, and psychological dimensions) dominated the course of human evolution for hundreds of thousands of years. In a very real sense our intellect, interests, emotions and basic social life are evolutionary products of the success of the hunting adaptation.

No inquiry into the social contract can be completed without review of what is known or speculated about man's hunting past. It is the period of human evolution that so decisively separated our way from the evolution of the largely vegetarian monkey or ape. If man's disposition for violence has evolutionary origins, then our long hunting history cannot be neglected through simplistic reference to the non-violent chimpanzee. If human social organization differs widely from that of the subhuman primate, then it is to our hunting days that we must look for possible causes. But there are many questions, and our authorities confront each other with opposed views. And so in this chapter I shall not only attempt to reconstruct our hunting way but to explore those arguments which would magnify or diminish its role in the human emergence.

While Washburn and Laughlin present much the same view, Washburn's is the more conservative statement since he refers to the hundreds of thousands of years in which hunting has dominated our course. The human brain experienced sudden enlargement a little over half a million years ago, and so Washburn refers to the time of true man. Earlier was the time of the hominid, the being who resembled ourselves in almost every way excepting his brain, a third as large. And Laughlin's reference to the transformation of a bipedal ape into a tool-making man is the larger statement, since it speaks not just of a character of man, but of the making of man. And this was the thesis originally projected by Raymond A. Dart, with his shocking paper in 1953, "The Predatory Transition from Ape to Man."

I told Dart's story in *African Genesis*—of his discovery in 1924 of the extinct small-brained South African being, *Australopithecus africanus*; of his claim that it was a hominid, a member of man's evolutionary line; of his conclusion that *africanus* was a carnivore, a hunter, and that he was armed with weapons long before the coming of man's big brain. The controversy was enormous and proceeds to this day. One question, however, that was only being opened when I wrote my earlier book has today been settled. In 1967 our foremost paleontologist, Alfred S. Romer, in the course of his presidential address told the American Association for the Advancement of Science: "Australopithecus and his kin are unquestionably morphologically antecedent to man, and with one or two exceptions all competent investigators in this field now agree that the australopithecines of the early Pleistocene are actual human ancestors."

In that most embattled and enchanting of scientific fields, where new discoveries compete with old prejudices to enliven every passing year, the slogan for professional and amateur alike must be "Publish and duck." One question has been settled. And another has been received in such scientific fashion that any may publish an answer without ducking. This question asks, How long ago? and a decisive contribution to study of the ancient past has come from the laboratories of the physical sciences. While the nuclear physicist has presented mankind with some unholy problems, he has presented anthropology with a variety of means for dating old wood, old lava beds, even old, elusive bones. Carbon fourteen, potassium-argon, fission track and other techniques all rest on the principle that certain unstable radiogenic particles will through atomic rearrangement achieve stability at given rates. Potassium, for example, has an unstable isotope which at an infinitely slow pace degenerates into argon, an element which in an ancient volcanic rock would not naturally appear. By delicate laboratory process the quantity of argon may be measured, so that we know with small possible error the rock's age. By such means do we announce the age of rocks brought back from the moon.

Absolute dating has in the past decade revolutionized our knowledge of the human past. For *African Genesis* my wife drew a chart of fossil men and what was known in 1961 of our ancestors. On it, as my best guess, she placed Dart's *africanus* as 800,000 years ago. Only an amateur with no reputation to lose would have dared a date quite so ancient. Yet now we know that the South African fossil bed can be no less than twice as old. And Clark Howell of the University of Chicago is today finding australopithecine remains in the beds of southern Ethiopia's Omo River twice as old as that. Kenneth Oakley's authoritative *Frameworks for Dating Fossil Man* gives the age of the Omo remains as about 3,500,000 years.

Much has happened in the decade since I was finishing the writing of *African Genesis* in the south of Spain. Two questions, then of utmost controversy, have been settled: the australopithecine ancestry of man, and the age in which these ancestors lived. Also, it is generally conceded that he ate meat. But in the best tradition of African anthropology, new discoveries have unloosed new storms of controversy, old prejudices have fallen into new fits of bad temper. New questions emerge, distinctly unsettled. How important to the survival of men or near-men was the eating of meat? Did we truly hunt, or did we scavenge the kills of others? If big-brained man for a half-million years killed animals larger than himself, did likewise his small-brained hominid ancestors? And if the hunting way has contributed as much to the human way as many authorities propose, then just how ancient is it?

The decade that has encouraged others to ask new questions has given me time, as well, to think things over. Whereas earlier I was preoccupied with the australopithecine's dependence on the weapon for survival, I find myself equally concerned today

with his dependence on society. Was it not the necessity for
early man, and earlier hominid, to hunt in cooperative bands
that gave us the evolutionary basis for present societies? And
was it not, within these societies, the division of labor between
the males who roamed in their hunting bands and the females
who stayed home with their slow-growing young—a pattern so
unlike most primate species—that set a social pattern which has
remained with us ever since? And would not these two characters of selective necessity—the cooperation of a small male
group and the separation but interdependence of the two divisions of the whole society—have placed a biological foundation beneath early consequences of varied sort, such as the need
for language?

It is an hypothesis useless to consider if no survival value was
attached to the eating of meat; or if we hunted such small animals that cooperative hunting was unnecessary; or if we merely
scavenged for a living, an activity little different from that of
the monkey or ape, scrounging on his own. Let us begin by inspecting the need for meat.

2

You will read many an informed article today stating that
throughout 99 percent of our history man was a hunter. While
the figure is conservative, it rests on most simple arithmetic.
About ten thousand years ago man began to gain control over
his food supply. Probably first in the Middle East we started to
domesticate such grains as wheat and barley, then a bit later
through selective breeding to domesticate animals like cattle
and sheep. Our attention turned to fields and to pastures. We
began to live in villages, then, with a surplus of food such as we
had never known, in towns. But the process was a slow one.
For several thousand years these early farmers still depended
for their food in large part on hunting. Not till five thousand
years ago could it be said that any significant portion of man-

kind had become independent of the hunting way. Yet if the hunting hypothesis as projected by Washburn and Laughlin and so many others is correct, throughout certainly the half-million-year history of true man we had depended on our skill as hunters for survival.

The way of life that we now take for granted and on the foundations of which we have built civilizations occupies but one percent of the time of the big-brain's preoccupation. And it is in this short period that present cultures have divided and subdivided, acquired peculiarities and parochial dedications. Earlier there was but one culture, that of the hunter, and its character was determined not by man alone but by the animals we hunted or competed with. A few of us, no doubt, depended on such uninspired activities as digging up clams, but such early peoples were exceptional. When the modern anthropologist asserts the essential unity of mankind as resting on the hunting way, he refers to the half-million years minus five thousand when, if our children were to eat, we killed animals large enough to feed us all.

The hypothesis is challenged not only by such reckless assertions as that of William and Claire Russell that hunting has absorbed us for a mere fifty thousand years, but by the far more respectable argument of many anthropologists that discounts the importance of meat in the diet of contemporary hunting peoples. The common phrase states that 80 percent of the foodstuffs in the diet of such primitive groups consists of wild plants collected usually by women and children. We were never, in other words, that dependent on the hunt. But there are at least two vital oversights.

Ethnology—the proper word to describe the scientific observation of primitive peoples—has a place in the reconstruction of our ancestral past, but its evidences must be handled with extreme care. That hominid and *Homo* included vegetable foods in their diet must have been as true then as it is today. But the vegetarian predilection of today's hunters encounters Richard Lee's summary of twenty-four hunting peoples in Africa and South America revealing only one in which as much as 80 percent of the diet comes from gathering wild plants. This is the Hazda of East Africa. But if we turn to James Woodburn's study of the Hazda, we shall discover that the percentage was calculated not calorically but by weight. There is a difference between a pound of lettuce and a pound of beef, if a man is to live out his years.

Even lacking the means to check the caloric validity of Lee's other cited sources, we find among the twenty-four warm-weather peoples in his list only five others who gain more than 60 percent of their food from plant sources. And when we turn to North American cold-weather peoples living above the latitude of New York City, such as the Eskimo or Chipawayan or Chinook, we discover just four out of thirty who look to wild plants for even 50 percent of their food. Yet these are the only modern hunting peoples whose protein necessity may be compared to early man surviving a half-million years of ice-age Europe and Asia.

Warm Africa, with lower protein necessity, was the scene of our still earlier hominid evolution, and Lee himself has contributed a carefully documented study that might seem to offer analogy. The !Kung Bushmen live at a series of waterholes in southwestern Africa. (The ! represents a language click which, after a decade and a half of frustrated trial, I do not recomment that the reader attempt.) Counting calories, not ounces, Lee found that 70 percent of !Kung diet is the product of gathering, not hunting. But what do they gather? In the favored area of this particular Bushman group is an ample forest yielding the high-protein mongongo nut. So hard-shelled is the nut that it may lie on the ground for a year without rotting. And the sensible Bushman prefers picking up nuts to pursuing animals all over the Kalahari Desert.

Lee's study is of an extraordinarily fortunate hunting people. Out of a daily intake of 2,140 calories, the !Kung Bushman derives precisely 190 from foods other than meat and nuts. That he thrives on such a diet would seem likely, and Lee found

10 percent of his population over sixty years old. But Henri Vallois of Paris' Musée de l'Homme, after a study of memorable dedication found that of 382 fossil remains of Pleistocene man, just ten had passed fifty years at death. "Few individuals passed forty years. . . . If the period from twenty to thirty years is considered as the main period of a couple's fecundity, it is evident that when the eldest members of a family reached adult age, usually their mother had already died and the father was not far from his end." Lee's fortunate !Kung Bushmen— an unfortunate choice for study if we are to look for revelations concerning pre-history—may with their lucky mongongo nuts have achieved a certain affluence, and even old age. Our ancestors did not.

Much may be learned from those living fossils, contemporary peoples who reject the modern way. That all possess a most exquisite knowledge of the natural world on which they depend speaks much for the human stock. That no living people, however primitive, speaks a language less complex grammatically than our own tells the linguist much about the probable antiquity of human communication. Such observations become even more dramatic when we recall that the few marginal hunting peoples remaining today are human evolution's losers, and that we are comparing them with those original winners whose genes were to seed the continents. But the traps facing the ethnologist are many. There is the second error.

Even were we to presume on the basis of contemporary evidence that our hunting ancestors were never in truth that dependent on meat, still the investigator would face a mighty fact. Our contemporary hunters are not that primitive. None lacks the use of fire or cooking utensils. And few are the natural nourishing foods other than meat—honey and nuts are examples— that may be digested by man without cooking. Yet fire and certainly cooking utensils must, from the long view of pre-history, be regarded as recent inventions.

Cereals, whether wild or domesticated, were unavailable to the human diet before we had a way to cook them. Crops and pots came as partners in the agricultural revolution, and that was a mere ten thousand years ago. So long as we had fire, roasting and baking were theoretically possible, and our use of fire boasts a longer history. Pekin man, 400,000 years ago, had his hearths at Choukoutien, and still earlier hearths have been found in Hungary and the south of France. All of the evidences of controlled fire lie in the north and date from periods of glacial cold.

Was fire used for cooking in such times, or simply for keeping warm? While surely man must have discovered that certain tubers and roots became edible if left in the heat of the hearth, and there are even the charred remains of hackberries from Choukoutien, still the evidence is quite simple that fire was for keeping warm. In Africa, a continent presenting many a problem but not that of freezing to death, we find no evidence for such early control of fire. Dart once believed that his Makapan Valley australopithecines in South Africa had used fire, but the claim has been disproved. Curiously enough, just a mile away in the same valley's Cave of Hearths we find sure evidence of controlled fire, as we do at Montagu Cave in South Africa's Cape Province, and at Kalambo Falls in Zambia. But none dates from over fifty thousand years ago.

If the conversion of vegetable matter into a fuel comprehensible to the human stomach was of survival necessity to evolving man, then it becomes impossible to explain why fire came so late to Africa. We must conclude, I believe, that the demand for vegetable foods on the part of our ancestors was marginal. Our own implacable children, confronting vegetables, reaffirm ancient decision.

Ethnology's analogy I believe quite false. Our remaining primitive hunting peoples depend more on the hunt than is generally accepted. And these people, all users of fire, have available far more wild vegetable foods than did our ancestors. Meat was never a luxury, as it is to the chimp or baboon. But we cannot leap to the conclusion that because the eating of meat was of survival value to evolving man, the cooperative hunting band was a social necessity. Small animals were available for traps and snares.

Few competent physical anthropologists would deny that in this last half-million years of big-brained man we hunted and killed large animals. Choukoutien, the site in China that I have mentioned, was discovered in the 1930's, and its caves show human occupancy from about 400,000 until 200,000 years ago. Pekin man, now classified as *Homo erectus*, was an early edition of *Homo sapiens*, and, like his Java cousin, *Pithecanthropus*, he had a brain about one-third smaller than average modern size. Yet in the very earliest of the Chinese deposits his remains are surrounded by masses of reindeer bones, evidently his chief item of diet. That his smallish brain of 1,000 cc. does not deny him the status of *Homo* may be judged by comparison with that of Anatole France, whose brain was precisely the same size. Though he wore the smallest hat in Paris, nothing prevented him from being one of the great French authors.

For two of Pekin man's contemporaries, Java man and a newly discovered skull from Tanzania's Olduvai Gorge, we have no direct evidence for hunting, but both give us precisely the same absolute date, 495,000 years ago. A remarkable site in Kenya, however, only a shade later, provides all the evidence one could need. Here we have no remains of the occupant, but we have his exquisite weapons and abundant remains of his prey.

Olorgesaillie was once a lake bed—or series of lake beds—deep in Kenya's Rift Valley, not an hour's drive from Nairobi. The site was discovered by L. S. B. Leakey, and elegantly developed and described by Glynn Isaac, who places its age as 400,000 years. (And in passing we should recall that this was 350,000 years before the first known use of fire in Africa.) Here for ages an ancient lake spread or retreated with shifts of rain and drought, and exposed or engulfed the record of those who lived beside it. We were true men without doubt; but whether we were of the early *erectus* edition such as Leakey found at Olduvai or of our own modern *sapiens* is doubtful. The quality of the weapons might indicate fully developed man, for they are the almond-shaped hand axes of the style known as Acheulian, beautifully wrought, man's principal weapon for hundreds of thousands of years. Their quantity, however, defies explanation.

The visitor to Olorgesaillie may today walk about on narrow, raised catwalks and observe for himself an outdoor museum of our ancient past. In one area twenty meters in diameter Isaac found 148 of the weapons, all from a single short period of occupancy. Another sixteen meters across produced 524. At an incredible third site of only 180 square meters, smaller than my apartment in Rome, Isaac found 620 weapons and tools all on a single surface. The observer looking down from the catwalk experiences the most eerie identification with the human past as he views several hand axes standing upright, points jabbed into the clay, just as they were abandoned by their owners 400,000 years ago.

And there were large animals in plenty, too. Their abundant fossils range from giant pig to horse to hippo. Most mystifying of the faunal remains left behind by these Acheulian men are the innumerable teeth of *Simopithecus*, an extinct baboon of the period. He was perhaps four times as large as the modern baboon, a creature who, as we have seen, gives enough trouble to contemporary man even with modern weapons. How did we with our hand-held weapons kill such primate monuments? One might conclude that it was impossible, and that we scavenged the kills of mightier predators. But where is the cat that could kill such baboons? There was the saber-tooth in those days, massive but unquestionably slow. That he did it is unlikely.

While I may suggest that none but the cooperating hunting band, working at night, could have disposed of *Simopithecus*, let us leave the question open until we have inspected the problem of scavenging in the next section of this chapter. For the moment let us accept that for the last half-million years true man has left evidence that he consumed animals far larger and more dangerous than himself. And let us turn our attention to the question of antiquity: Is it only in the time of *Homo* that we find our fossil remains associated with the remnants of animals larger than ourselves, and that we could have killed—if indeed we ourselves killed them—only through the cooperative efforts of a band? Let us move back a few paces in time.

In *African Genesis* I described Tanzania's remote, dry Olduvai Gorge as the Grand Canyon of Human Evolution. And through the devotion of Louis and Mary Leakey it has since proved to be just that; tourist buses run to it today. By 1961 the Leakeys had found, beyond an immense number of stone artifacts, just one fossil hominid, which Leakey described as *Zinjanthropus*, the Nutcracker man. I enhanced, to my regret, no subsequent cordial receptions on the part of Dr. Leakey by correctly describing it as a variant of *Australopithecus robustus*, a vegetarian cousin of Dart's southern *africanus*, discovered by Robert Broom in South Africa many years before. *Zinjanthropus* is today so classified. *Robustus*, with his huge molars specialized for the munching, largely vegetarian life, and the crest on the skull to which necessary, powerful muscles were attached, could never have become a human ancestor. He was a side line.

Then however, several years after the publication of my book, Leakey announced the discovery of what he called *Homo habilis*. Here was the meat-eater with the slim, light teeth, the clean skull undecorated by crest, resembling Dart's *africanus*. His brain was a shade larger, his anatomical characters were closer to the human line. Here was a being as close to the veritable human ancestor as one was ever likely to find. And the Leakeys had found him at the bottom of the Olduvai Gorge, in what is known as Bed One, with an age confirmed by both potassium-argon and fission-track dating as almost two million years. *Habilis* was a hominid, four times as old as our most ancient traces of true man, with a brain less than half the size of ours. The shock of the discovery, spreading out from Olduvai's dusty range, provided an earthquake for anthropology from which it has never recovered.

Arguments skittered about like cats on a hot scientific roof. Leakey denied that *habilis* was an australopithecine. Today almost every authority—Le Gros Clark, Oakley, Campbell, Robinson, Howell—disagrees with him. *Habilis* was a halfway house on the evolutionary slopes of the climb to man. Whether we call him *Homo* or not is a matter of semantics, so long as we do not deny that he was an advanced member of Dart's *africanus* family. But whether or not he was a successful hunter of large animals is a question far from semantics. If he captured nothing but turtles, snared nothing but rodents and birds, seized as do modern baboons nothing but the gazelle's helpless fawns, then the hunting band was unnecessary. The social organization of ancient, small-brained hominids of the human line rests in part on the answer.

We know today that all of Bed One, Olduvai's oldest deposit, and the lower portion of higher Bed Two antedate by at least a million years our earliest evidence of true, large-brained man. Whatever is found in these deposits, accumulated ever so slowly beside a forgotten lake, has almost all been derived from the hominid living-sites excavated by the Leakeys. The stone implements were the work of a being with a brain less than half the size of ours, and Mary Leakey's painstaking analysis reveals them not just as crudely chipped pebbles but as a true stone industry, a variety of tools or weapons following specialized styles. As a shaper of stone, *habilis* was no novice. Was he then a novice hunter?

In the confusing early announcements of the *habilis* discovery, sensationally but inadequately reported by the world press, he was so described. And in many a mind that should know better the impression still lingers: before true men the hominid killed only small or very young animals. But in the years that followed the discovery, fossil remains of prey animals littering the living-sites have been identified of a size to challenge the lion. That they were eaten finds its evidence in the quantities of bone smashed by stones to extract the marrow.

A large extinct waterbuck, for example, appears at the same level as the earliest site. Another, an extinct antelope called *Strepticeros maryanus,* named for Mrs. Leakey, was first found in Bed Two, then later in number in early Bed One. It is related to the kudu, a favorite today of big-game hunters. The eland, largest of antelopes though not the most dangerous, has yet to appear in any early sites. But *Hippotragus gigas* does. Leakey describes it as a creature "of gigantic size," larger than today's sable and roan antelopes.

I find most impressive Leakey's description of a large, edible, extinct beast found on the earliest *habilis* living-floor. The animal has been classified as *Oryx sp indet,* which means that, while it was a member of the oryx family, no one can make up his mind which species it was. Leakey describes it as "more massive than any living species of oryx," and if he is correct, then lions should flee.

The most massive oryx today is the gemsbok of South West Africa. The genus is characterized by long, straight, lethal horns which, unlike most antelopes, they use not just for display but to defend themselves. George Schaller has found that East African lions almost always kill by suffocation. But the lions of southwestern Africa's Etosha Pan have learned better. To grasp a gemsbok by the throat accepts the probability that the straight, sweeping rapier horns will disembowel you. And so the local lions have developed a safer tradition. They take the gemsbok from the rear, break its back at a vulnerable point, and go off to sleep until the animal dies. Yet *Oryx sp indet* was larger.

Nothing in *habilis'* faunal assemblage—the variety of animal companions found with his remains—confirms the belief that he confined his butchery to turtles and rodents and birds. In a later year the Cambridge University Press published a definitive study. A specialist, L'Abbé Lavocat, considered rodents and found that at the early *habilis* sites their number varied between few and very few. Only at a site at the top of Bed One

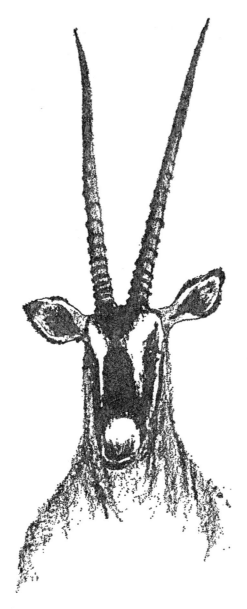

and the bottom of Bed Two, some hundreds of thousands of years later, did rodents appear in number. Dr. Leakey in the same publication analyzed the remains of *Bovidae*, which at

Olduvai we may loosely classify as antelope. He commented on the abundance of such remains, and the taste for such grazers and browsers which man has always had. (At Choukoutien, reindeer contributed three quarters of all animal bones.) Apparently the near-man of Olduvai Bed One had the same taste. Dr. Leakey wrote:

> Perhaps the most remarkable fact that emerges so far is the scarcity of fossils representing the smaller members of the Bovidae, for example the duikers, the dik-dik, the oribi, the steinbok and klipspringer. This cannot be due to the bones of these species escaping notice since thousands of bones of much smaller animals are in the collections. There must therefore be some other and at present unexplained reason for the scarcity of small antelope remains on the living-floors of the early hominids.

Leakey's comment, which closes his long section on the Olduvai mammals, suggests to me that the great scientist, who has contributed more material to our understanding of human evolution than any other man in history, apprehended more than he cared at this date to commit to record. The absence of small antelopes would affirm the reality of the hominid hunting band.

Only solitary hunters—the leopard, or the cheetah skulking long-legged through concealing grass, flat diamond-shaped snake-like head held low and inconspicuous until its range has been closed and it can unleash its blinding speed—regularly kill prey smaller than themselves. The solitary male has only himself to feed, the solitary female herself and her cubs. A duiker will do. But social hunters have many mouths to feed, and small prey become a mouthful apiece. So the wolf pack takes caribou and moose, the hyena takes wildebeest and zebra, the lion pride accepts the lethal risks of the African buffalo. When times are hard, of course, they will take what they can get. Wolves will eat mice. But it is not their way. Cooperative hunting makes possible and demands the big kill. Only the big kill makes possible the appeasement of many appetites.

Four times as long ago as our earliest example of large-brained man, a hominid weighing eighty pounds with a brain less than half the size of our own killed prey animals as formidable as any we know today. Possessing only such hand-held weapons as could kill at close quarters—wooden or bone bludgeons, stones well but more crudely chipped than those at Olorgesaillie—the feat would have been impossible had *habilis* and other australopithecines not hunted in highly organized, cooperating bands. Even with such bands the difficulty seems such that we must take seriously the proposition that he did not himself do the killing but scavenged the kills of others.

Let us inspect the predatory community, which I believe will provide the answer.

3

The Gorongosa is an animal Eden lying low near the southeast African coast where the warm Indian Ocean breathes annually its damp monsoon on the sprawling land. And so, unlike most African game reserves—dry, high-altitude, open—the Gorongosa is a tropical garden. Palms finger its horizons, palmetto thickets conceal waterbuck rumps. Flooding rivers in the wet season leave long pastures in the dry. Rain forests with drooping lianas surround glades supplying hartebeest comfort. Unending groves of fever trees satisfy the visitor with a strange lime-greenish world, satisfy the leopard with ample crotches to climb to, satisfy the happy elephant with more practical-sized trees to push over than he can ever get around to, even if he lives to a hundred. And there is so much to eat. Never have zebras been sleeker, hippos louder, buffalos so stupid, baboons so numerous, sable antelopes so visible, lions so pleased with it all. But perhaps the best thing about the Gorongosa, from an animal viewpoint, is that until a very few centuries ago men never came here at all.

Not many men come here even today. Mozambique—southern Africans usually speak of it as Portuguese East—is a land somewhat the shape of California and Oregon together, somewhat larger, and somewhat more difficult to get to. Its giant park, stretching back to the mountains, lies a hundred miles inland from the port of Beira, but how to get to Beira is a problem. If one reaches it from Lourenço Marques, hundreds of miles to the south, then how to get to Lourenço Marques becomes the problem. When the persistent collector of game reserves, however, at last reaches the tidy lodge with its cabins and rondavels, its lawns and encircling flamboyant trees, its swimming pools for old and young and even its African attendants for children whose parents go lion-watching, he will find only eighty beds. He will encounter no traffic jams along the tracks and trails, just as the lion whom he encounters will regard him as a rare, unimportant, and most definitely boring being.

If visitors to the Gorongosa have it good, then so have the lions. There is so much game that one need scarcely stretch out a paw. One singular pride has achieved an affluence available nowhere else on the continent. Years ago the Portuguese, a bit naïve in such matters, built a group of tourist cabins too close to the summer floods. The cabins in the end were abandoned, and lions took over. Today a lioness looks out a window, yawns in the midday heat; the tail of some sleeper hangs over the edge of what was once the dining-room roof. Your car passes. The lioness in the window, like a grandmother in an Italian hilltown, yawns again. She was not that interested in your passing in the first place.

It is a leisurely, ordered, luxurious world bearing strong resemblance, perhaps, to that other Eden in Miocene East Africa, twenty million years ago, where the first human ancestor took the irreversible decision to become something more than an ape. In those early days the East African rains were heavier, the rivers brighter, the forests denser, the pastures richer. Like the Gorongosa, it was an animal paradise endowed with all things but men. And in such a setting did we, the infant hominid, on the path to bipedalism take our first step.

The visitor to the Gorongosa—or to any African game reserve clutches invariably a hope that he will see a lion make a kill. The hope is a comment on our hunting past. Be the visitor clergyman or soldier, housewife or schoolmarm, Japanese or Swede, it will be always the same. Jung wrote: "Just as our bodies still keep the reminders of old functions and conditions in many old-fashioned organs, so our minds too, which apparently have outgrown those archaic tendencies, nevertheless bear the marks of the evolution passed through, and the very ancient re-echoes, at least dreamily, in fantasies." The visitor's fantasy is an echo of our universal human experience.

Man identifies himself with the lion. And that is why the lion is the most important of animals in the human view. Chimps, rats, rhesus monkeys may adorn our laboratories. It is the lion that adorns our coins, our gates, our museum steps, our cinema trade-marks, our coats-of-arms, our metaphors, our memories of things we never saw. His cerebral capacities may be unworthy of mention; we like his mane. His indolence would affront the tortoise; we envy his roar. The male's despotism would make a Stalin shudder; yet the most idealistic student, entering a game reserve, will hope to see a lion make a kill. It is the way we are, and the way the lion is. Even when with a certain sophistication we recognize that the vast male is too lazy to kill and will almost surely leave it to the ladies, still with unwavering fidelity we shall treasure the image engraved in olden days on our collective memories: Lions are power. It is Adler plus Jung.

The image has good reason, for this is the beast we faced before the advent of weapons that could mysteriously kill at a distance. There is a myth in science, that the lion fears fire. The myth was punctured by George Adamson. Few men on earth know so much about animals as Adamson, senior game warden for a generation in Kenya's wild Northern Federated

District. Though man-eating lions are rare, a portion of his duties was to find and kill such beasts when from some native village reports came in of excessive human loss. In a Nairobi hotel room I asked, "But how would you know which was the man-eater?"

It is really quite simple, Adamson told me. You have only to go to the village, retire to its outskirts, and light a fire at night. There you sit, your gun across your knees, and you wait. The lion that is not a man-eater will stay away, since the fire is a signal of human presence. But for the same reason the man-eater will come. And so when you see eyes approaching, you shoot.

I myself do not recommend the Adamson system of human bait as a way of life. But his experience demonstrates that it is not fire that the predator fears but the man beside it. Like most animals, he has acquired a respect, by now probably innate, for man and his weapons that kill at a distance. But if through some chance he has tasted human meat and the taste has been attractive enough to overcome his fear, then fire will not repel but attract him. That fire in itself will frighten off predators is an anthropological myth. And we may even speculate as to why the use of fire in Africa, with its vast population of great cats, was so long delayed. Before we possessed such a long-distance weapon as the bow and arrow, they had no reason to fear us or to regard us as other than prey. In such a time and place fire was no friend.

For good reason we fear and adore the lion, and call him king of the beasts. We bested him with our long-distance weapons, and so today in a protected area he ignores us as a sweaty *nouveau*. But if we were hunters, then for millions of years we were members of the same predatory community. Within the large assemblage of species we were fellows and competitors in killing for a living. And in our predator community—cheetah, leopard, hunting dog, wolf, hyena, the long-gone saber-tooth cat—the lion was king.

Our newly acquired knowledge of the predator community

will clear up many a question, contradict many an assumption. Until today we have known almost nothing about the more dangerous predators, and our ignorance has been as great a handicap as was a decade ago our ignorance of the primate. But the long, definitive studies by George B. Schaller and Hans Kruuk, although they may not yet have reached press when my own investigation is published, provide the evidence we need. And a principal conclusion concerning any natural association of predators is that all species kill and all species scavenge. The intensity of competition between predators to scavenge any kill is of an order that could have left small room for that ill-adapted little carnivore, our ancestor.

The key to the conclusion was Kruuk's observation of just how formidable a hunter that reputed coward, the hyena, can be. Kruuk began his studies in 1964 in Tanzania's Ngorongoro crater. Two million years ago, when australopithecines frequented the nearby Olduvai Gorge, Ngorongoro was an immense volcano. Subsequently it collapsed, leaving in its interior what geologists describe as a caldera of one hundred square miles. With year-around water and year-around pasture and forage the caldera became another animal Eden which has attracted one of the densest populations of hyena in all Africa.

Kruuk recognized that there were not enough lions in the area to provide meat supply for themselves and the hyenas too. The hyena packs must be killing on their own. Yet there was little sign of such activity. When did they do it?

Kruuk solved the problem by following packs at night, driving cross-country without lights. How he solved the problem of not killing himself, I do not know. But the hyena secret was revealed. The animal has most sensitive nocturnal vision, far more so than have most prey. Zebra or wildebeest that could outrun the pack in the daytime cannot at night. And so the hyena by concerted action can catch and kill virtually anything. In a thousand observations of hyenas feeding, both in Ngorongoro and on the Serengeti plain, Kruuk found that on over 80 percent of occasions they were eating prey that they had killed themselves.

The hyena, however, has an emotional problem. The excitement of the kill is just too much for him and so the pack sets up a racket to grace an inferno. Lions for miles around are notified that a free meal is available. Kruuk has made a tape recording of the din, and, played anywhere, it is guaranteed to attract lions within fifteen minutes. In the Ngorongoro, lions hardly ever bother to hunt, and with their usual dim view of hard work live in affluence off hyena effort. Complementing Kruuk's observations, Schaller has found that in the Serengeti, where hyenas are fewer and lions must work harder, still about 25 percent of what seem lion kills have been killed by somebody else. Hyenas, of course, are formidable enough so that a pack can resist two or three lions, but, reinforced by a friend or two, then the lions will take possession. And in the meantime the leopard, out of deference to the overwhelming larceny of both lions and hyenas, has evolved a normal behavior pattern of taking his prey up a tree to safety.

Some years ago De Vore and Washburn published their opinion that evidence could not support the scavenging hypothesis, and that scavenging could have become a source of meat only when man became a hunter formidable enough to drive other carnivores off a kill. What they wrote of man could only have been more true of the smaller, more poorly armed hominid. It is a view that Schaller, on the basis of the new evidence, supports. In 1969, fascinated by the problem, he and a colleague, Gordon Lowther, turned themselves into hominids and, unarmed and on foot, went out into the Serengeti to see if they could make a living. They caught a sick, abandoned zebra foal and a young giraffe that behaved strangely. Captured, it turned out to be blind. They concluded that quite primitive hunters could have gained a fair meat supply from such prey. And indeed, as we have seen, the weeding out of defectives is a normal function of predators. But although it was a period of birth-peak among gazelles, they had the opportunity to take few fawns and dismissed them as a significant source of food. So far as scavenging was concerned, by following the flight of vultures they came on a buffalo that had evidently died of disease or old age. Despite earlier scavenging, a fair amount of meat was still on the carcass, but Schaller records that the discovery was a matter of luck and could not happen frequently. The remains of lion kills offered nothing but a few pounds of brains protected by skulls that neither lions nor hyenas had been able to open. Schaller concluded that so slim and unreliable were the rewards of scavenging that only by persistent and cooperative hunting could the hominid have commanded a reliable food supply.

So long as we accepted the hyena in the cowardly image that has prevailed since the days of Herodotus, then we might be free to visualize our meat-eating ancestors as driving off the craven pack while they themselves made off with the leftovers of lion satiation. With our new information, however, we must conclude that the hyena, quite capable of killing adult wildebeest, would have been only too happy to dispatch us.

The predatory community is a scavenging community as well. The leopard has excellent reason to take his prey up a tree, for he cannot compete; and the hominid was less formidable than the leopard. Despite the ferocious competitions, we may with luck in our hominid days have made off with a stolen joint or two, as Schaller has demonstrated. But the bone piles adorning ancient living-sites were in large part the rewards of hunting. Smarter than hyenas, we did not announce a triumph to our competitors, but sneaked off with our prey and kept our mouths shut.

The cooperative hunting band was a reality in the lives of true men and the earlier hominids wherever bones have collected on living-sites. We could not otherwise have killed as large animals as are found, and scavenging could have accounted for no significant fraction. The band was of survival value in defense as well, for we were edible. As an evolved primate, we were as attractive a prey as the baboon, and we lacked his capacity to seek rapid safety in trees. And so we faced two ways: we hunted and we were hunted.

I shall waste no reader's time on the quaint proposition, advanced by some, that before we had adequate weapons the great carnivores feared us as they do today. There is a remarkable study, however, published in March 1970, proving that carnivores dined on australopithecines. It is worth recording for its scientific ingenuity alone.

When C. K. Brain returned from Rhodesia to South Africa's Transvaal Museum, he determined to reinvestigate an australopithecine site called Swartkrans, about an hour's drive north of Johannesburg. It was a site discovered by Robert Broom. Unlike Raymond Dart's great Makapan cave, which had been occupied solely by *africanus*, Swartkrans revealed only the remains of *robustus*. A lime deposit in the very high veld, Swartkrans has always had an enigmatic character. Broom and later John Robinson found scores of *robustus* individuals together with the remains of thousands of animals. But if *robustus* was largely a vegetarian, as Robinson demonstrated, why should there have been such butchery? Then a good many years ago a toothpaste manufacturer seeking lime blasted the deposit to pieces, making it so enigmatic that the scientists gave up. And this was the mystery that Brain determined to make sense of.

And sense he has made. The site was never a cave but a natural drain carrying surface water underground. In this pipe-like formation was deposited lime enclosing the bones washed down from above. The analysis and classification of fourteen thousand fossils by Brain and his talented wife showed a definite limitation of animal size. With few exceptions, nothing like the larger antelopes or zebra appeared. *Robustus* was about as heavy as they came. Brain next went into the Kruger Park and analyzed the evidence of leopard kills. The range of size was the same. The limit was the weight that a leopard could take up a tree. *Robustus* had been larger and stronger than *africanus*, and likewise armed to defend himself, as his small canines attest. But he had been as vulnerable to the leopard as wart hog or springbok or any prey of proper size.

There remained a problem. Why should leopards always have come here? Brain studied the nearly treeless high veld, consulted botanists. There is a rare clump of trees beside Swartkrans' limy pipe, encouraged by the drainage of water. Trees had always been there. Leopards, to escape scavenging competition, had for tens of thousands of years made for the rare trees with their prey. There in a crotch they ate their fill, letting portions drop below, later to be washed down the drain and to create future mysteries. They ate the hominid in plenty, but they ate true man as well. There is at least one *Homo erectus* in the assemblage.

This is the essence of as elegant a reconstruction of the past as paleontology can provide. It proves, above all, our vulnerability in the time of the hominid coming, and the proposition that we faced two ways: those whom we would kill, those who would kill us. An objection may be made that the vegetarian *robustus* had a better flavor than the meat-eater, *africanus*. But Kruuk has the record of an Ngorongoro leopard who dined almost exclusively off jackals. And the objection, too, overlooks the hyena (there were more species then than now), more formidable than the leopard, who will eat anything, including his brethren. That the vulnerable way was ours is the most haunting conclusion of Brain's study. There is a by-product, however, as striking.

The difference between the Swartkrans fossil assemblage produced in large part by a solitary killer and the fossil assemblage associated with *africanus* at Makapan is total. Seventy or eighty thousand Makapan fossils have by now been developed and classified by Dart, James Kitching, and their crew. Antelope of medium or large size—all beyond leopard capacity—dominate the deposit. Far more thoroughly investigated than Olduvai's *habilis* remains, the fossils follow the same pattern and point to the conclusion that in more primitive *africanus* days we were the same social hunters with the same hunting bands who, despite all danger, favored large prey as our only source of sufficient food.

We were lions among lions, though, vulnerable as we were, we could scarcely have lolled in such indolence. As with the lion, however, courage and cooperation were the heralds of our future kingship. And there was something more still. During a short recent stay in the Gorongosa, I found that on two nights lions killed hippopotamus. Only a decade or two ago it was regarded as an impossible feat. Admittedly the Gorongosa lions are enormous, as their prides are enormous; some can field a hunting team of a dozen adult and sub-adult killers. But the 5,500-pound hippo is likewise enormous and with jaws like an armored steamshovel can crush a lion as a dog snaps a kitten. Grazing far from his protective pool, he is as dangerous a prey as imagination affords. One knows how the lions did it: through the same tactics and cooperation as we used in our hunting days. But legitimately we may ask why did they do it, with such an abundance of large and less menacing game at hand?

We may ask the same question about *habilis*—lacking fangs, claws, proper physical power—as beside an ancient East African lake he faced *Oryx sp indet*. And if ever we come on a reliable answer, we shall know a fortune more about lions and men.

4

When our years proceeded like dusty moments and millennia were but long drawn days, and even a million years bore no punctuation marks but the slow cycles of prevalent rains, prevalent droughts, spreading forests or eroding sands, then natural selection with divine unsentimentality encouraged or discouraged the being that would eventually be man. It was a time for imperceptible failure, unappreciated success. We lived. We drew necessary breath. We did our best according to biological command as old as the amoeba. We reared our young as does the elephant because they were there. We greeted life as the lark greets the morning, for no excellent reason. We died reluctantly, as dies the wart hog, for no reason more appreciable. We climbed the mountain. Why? We languished in the valley. Why? We did our best, and, doing our best, we obeyed the same laws as the lion or the monkey, the Uganda kob or the slim impala.

The ancestral hominid, lacking significant brain or other imposing assets of body, surrendered himself without protest to natural dispensations. He lived, died, loved, lost, rose, knelt, fought without mercy, embraced without qualification, suspected, accepted, conducted the unconscious human pilgrimage within the stringent boundaries that survival laid upon him. That he produced us was no part of his plan. Yet as we look back at his ancient trail we must accept him as a part of ourselves. If we have foresight, then his were the necessities that lent foresight selective value. If we are self-aware, then his were the glimmerings, the wonderings, the vague intimations that fingered the way. If we can speak, then with little doubt the beginnings of language were his. But to consider only the beginnings and the gropings is not enough. For he perfected a social group more reliable than the vertebrate world had ever seen, or would ever see again. So strong must have been its ties, so settled its number, that its character remains with us today.

The hunting band we may calculate as a group of nine, ten, perhaps eleven adult and sub-adult males, supporting a whole society of about fifty. Contemporary hunting peoples live in smaller societies of about twenty-five, but with their weapons they may kill at a distance and the hunters have less need for each other. Very little longer than ten thousand years ago,

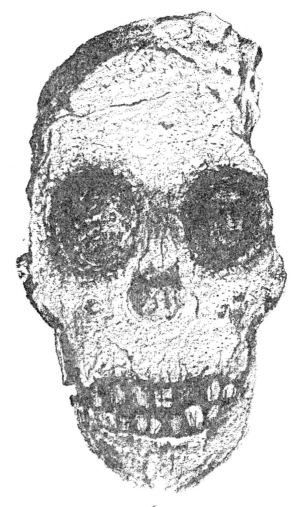

whether true man or hominid, we killed at close quarters. The wooden club, the heavy bone, the hand ax, the thrown stone were our only decisive weapons. Surely we made traps, arranged snares. At a Solutrean site in France, Neanderthals drove wild horses over a cliff, perhaps for millennia; the fossilized remains of thirty thousand horses lie at the base. For the normal hunt, however, too few hunters would have been unable to surround or ambush the game. And too many would have implied a community of women and children bigger than a hunting range could support.

The band itself was almost surely all-male. In the lion pride the male, not only lazy but heavy on his feet, leaves the hunting indolently to the lionesses. Earlier I mentioned the peculiarly reversed sexual dimorphism of the hyena. Since the female is larger and stronger, hyena hunting packs are mixed. So likewise mixed are the other major social hunters, the wolf and the African hunting dog. But the hominid, an evolved primate, suffered the misfortune that nature did not design him to be a hunter. We lacked not only power, speed, and lethal accessories: we lacked, as do all apes, children who would grow up in a hurry. And so probably from the beginning of meat-eating our female ancestor found that her place was in the home. The young lion may by an age of less than a year have achieved such independence that he needs little adult attention. Hunting-dog pups at six months, as I have described, can very nearly keep up with the running pack. But the young of no higher primate, chimp, gorilla or hominid, conveys such privilege on the mother.

That in our evolving days the female did not hunt has suggested to Adriaan Kortlandt a most curious hangover virtually universal in our species. It is the "girlish" throw. Women throw underhand as does the chimp, and few with whatever frustrating practice can achieve the overhand throw of the man. His is a motor pattern which only the male has inherited, perfected by many thousands of generations of armed hunting.

Our hominid social group suffered, in consequence, functional segregation. There was the band of adult and sub-adult males who went out on the hunt. And there was the home group at cave or living-site, the women, infants, boys too young to hunt, girls too young to reproduce. These were the collectors who scoured the locality for scarce edible fruits and plants, snared a bird or rabbit, even caught a fawn or two. And so there was the man's world, and there was the woman's world, and there was obligation. Of almost two hundred species of living primates, we are the only one in which beyond the time of weaning anyone feeds anybody else. With meat-eating came division of labor and the obligation of the hunter to feed those who could not hunt.

The sexual segregation of adults in the hunting society marked the beginning of human division of labor. When pups are small the hunting dog, gorging himself at the kill, brings home food that he regurgitates not only for the pups but for those who have stayed home to guard them. This too heralds division of labor. But the guards may or may not be the mothers, or even females. It is a constantly shifting arrangement. In hominid society began the first permanent social division based on functional contribution to the society as a whole. And with it came interdependence, and compulsory responsibility. If the hunting band forgot its obligations and failed to return from some savanna saloon, the society would soon perish. Natural selection, operating at the level of groups, determined superior social responsibility as a condition of survival.

Social obligation was born on the African savanna long before the coming of the enlarged brain. The hunting band killed for the group, not just for itself. And the daily or frequent necessity to bring back the meat to the living-place brought other determinants to the hunting life. Leaving meat on the savanna after a kill was to surrender it not only to our competitors in the predatory community but to the omnipresent, all-seeing vultures as well. It was this necessity to return that set a limit on the size of the hunting range, which in turn set a limit on the size of the total group. And the limit of range was probably smaller in hominid days than in the time of true man. John

Napier has made a remarkable study of the foot bones of *habilis* and concluded that at this stage of our evolution we could run well enough but lacked the capacity for the walking stride. Long marches were denied us or made difficult.

It may have been this restriction on the size of our hunting range that introduced the territorial principle of fixed boundaries and exclusive use of space to the hunting primate. Territory, as we have seen, appears sporadically in the vegetarian primate. Definitely it is a portion of the primate behavioral repertory, but just as definitely it is not always expressed. For the social hunter, handicapped by an incapacity to roam far distances, a protected and exclusively possessed hunting territory was compulsory, or rival hominid bands would make off with his game. Even the highly mobile wolf, lion, and hyena divide up potential hunting space. Only the hunting-dog pack, so fearsome that it drives all game from an area, is non-territorial and must keep moving to fresh areas, keeping ahead of its reputation.

We were not so fearsome. And the earlier we go back into our evolutionary times of anatomical restriction, the smaller our range must have been and the more severe our need for exclusive space. Territorial behavior, I should surmise, was an early imperative in the lives of our ancestral hunters. But I should likewise speculate that until expanding populations enhanced demand for the best hunting space, competition was not too severe. We practiced, like many a species, territorial spacing through avoidance. We had disputes and feuds, undoubtedly. But we had troubles enough, all of us, to discourage other than occasional shouting matches with troublesome neighbors.

It was communication not with neighbors but with one another which must interest us, for cooperative hunting would in itself have encouraged the development of rudimentary speech. Of our predatory competitors the lion has developed tactical hunting beyond any other. A group of lionesses will size up the prey and its situation, take advantage of any obstacle like a stream or bog, leave one or two, perhaps, to attract attention while the others vanish to perfect the ambush, then, as a trap is sprung, drive the prey directly into the grasp of hidden killers. George Schaller has furnished me with a series of sketches drawn from his observations, and they resemble nothing so much as plays in American football. Yet lions fail frequently through imperfect timing. Hidden from each other, someone fails to reach position when the trap is sprung.

A capacity for vocal signals and warnings would have been of selective benefit to any hominid hunting band. Yet I doubt that the hunt itself was the cradle of language. Signals little more elaborate than those of the baboon or rhesus monkey would have provided an ample repertory. Where true language would contribute survival value, in my opinion, lies in the storage of information as a social asset of the hunting band, and in the education of the young.

I have discussed in an earlier chapter the function of education as second only to defense in most animal societies. And the slower the maturing of young in a given species, the greater therefore must be the demand for maximum learning. Slow growth would otherwise be a selective disadvantage. We must assume, then, that education rated high in hominid social values. Yet the hominid in his vulnerability was an odd sort of hunting creature. You did not take your little boy along when you went hunting *Oryx sp indet.* How, then, did he learn to hunt?

Let us reflect again on the segregated nature of the whole hunting society. The home was a definite concept, as the restricted size of bone-strewn, tool-scattered living-sites attests. At the two-million-year level of Olduvai the Leakeys found an oval construction of stones which they interpreted as the foundation for a rough shelter. I remain doubtful. In 1969 such oval constructions, definitely foundations, were found in the south of France just a few hundred yards from the harbor in Nice. But just as definitely they are a mere three hundred thousand years old, and their remains are unique until relatively recent times. The Leakey discovery, I believe, remains one of our better anthropological mysteries. There was little need for shelter in the Olduvai Gorge, and there is the problem of movement as well.

The existence of living-sites does not imply year-round occupation. Seasonal movement along with the seasonal movement of game must have occurred. Territories themselves may have shifted. Still, wherever the society camped, deeply engraved on our evolving minds must have been two ideas: There was the home-place where women put down their infants, where young-of-an-age formed peer groups and played at the hunting game, where men came home weary with their loads of meat, fat or slim as luck dictated, partially butchered probably at the kill. It was as if we acted out two charades, and the home was one stage. The other was where the action was, the wide world of the hunter, the overbearing, menacing bush, the suspenseful waterhole, the limitless savanna defying and challenging the evolving foot. The hunting band's scene was one of alertness, of discovery, of plans, of lifetime camaraderie, of violent action, of dawn-to-dusk danger.

The bipolar nature of hominid society—the functional segregation, the physical separation, the disparity of styles, routines, and goals—I suggest became the cradle of language as we know it. Things had to be told. A hunter was injured, a child was sick. Hunters returning empty-handed had to tell apologetically of the big one that got away. Leopards menacing the home-place had to be described by the women, numbered, placed on the map if the group was to be defended. As important as any motive, granted the commands of hierarchy and alphaness, was the hunter's necessity to boast before an audience when the big one did not get away; to tell and retell the precise details, the most remote contingencies of the day's heroic deeds. We still do it. Thus, to the selective advantage of the group, the experiences of the day's hunt became committed to the traditions, to the lore, to the wisdom and advantage of future generations. The boys were listening.

Perhaps it is a crude way to state it, but the difference between animal and human languages is storytelling. Today's explorers of linguistic frontiers, men like Noam Chomsky, Eric Lenneberg, Thomas Sebeok, find impossible the child's rapid learning of language entirely by associative learning, by reinforcement theory—in other words, by an effort to please the parent or to avoid his displeasure. Most parents I am sure will agree. A biological basis for the learning of language demonstrates itself in the learning of grammar. Some inborn pattern must exist to dictate not just the learning of words but their proper sequence. I myself see the origin of grammar in storytelling, in subjects and predicates, in tales told so long ago that they preceded the expansion of the human brain and contributed to its organization.

Animal language describes or defines a situation as it exists: to alert, to warn, to ring the bell of action or the bells of rejoicing, to beg, threaten, appease. Human language recalls a situation, relates cause and consequence, anticipates the future. It tells a story, points a moral. The hominid hunting band that even on a most rudimentary level commanded such an instrument scored a selective triumph over its predatory competitors. Information could not only be stored but, like the rule of kings or the wealth of merchants, be passed on to descendant generations. For the boys, as I have said, were listening. And slowly was created the shining weapon of human communication, and the means of preparing the immature and the vulnerable, through the sophistication of listening, for the day of the buffalo and the night of the lion.

Looking back across that incalculable sea from the time that is now to the times that once were, I find it beyond our powers of discrimination to evaluate like some deceased's estate our varied inheritances from the hominid. Which was worth more, which less? Which has enhanced the hopes of civilized man, which, maladaptive, has damaged us? For certainly two million years we were continuously dependent on the weapon in the hand to make possible the survival of a terrestrial primate so ill-armed by nature. Without the invention of the weapon, we could not exist. Yet our affinity for a cultural device that made our existence possible became, as we all know, a dubious legacy. But through the same unimaginable time we were perfecting cooperation, social obligation, individual responsibility to the group at a level to which no other primate had ever risen.

It is as if certain patterns, certain biological short cuts—I call them propensities—perfected or pioneered by the early hominid contributed circuits to the nine billion neurons of the enlarging brain to come. Thus the storytelling patterns may have supplied a biological foundation for what we call grammar. And as the weapon has thrown deep shadows across human history, so language has brought it what light we know.

In its influence on the modern brain, less immediately apparent is the nature of the all-male hunting band. I have described it as a cooperating group of nine or ten or eleven able-bodied adults and adolescents. Smaller groups undoubtedly existed, but at disadvantage in the cooperative hunt. And a number much larger would have involved, as I have said, a community of dependents too numerous for the limited hunting range to supply. In the very earliest days, of course, the African savanna must have been, in Kortlandt's phrase, a butcher's paradise. Unskilled as we were at the hunt, prey animals granted us but passing fear. Flight distance must have been slight, and our problems of stalking negligible. But vividly we must also recall that in such times, long before *habilis*, our feet were yet poorly adapted, our new bipedal posture ungainly, our weapons awkward and grasped by inexperienced hands. The naïveté of the hunted was balanced by the naïveté of the hunter. We must likewise recall that an injured animal is an infuriated animal. And our powers of flight, in such an era, must indeed have been lamentable.

I can visualize no time in the history of the meat-eating hominid when hunting was other than a dangerous trade. As our skills and our anatomical adaptations evolved, and our reputation on the savanna worsened, so evolved likewise the wariness and the defenses of our prey. That we succeeded was a tribute to native primate wit that neither prey nor our predatory competitors could match. If a wolf pack, hunting cooperatively on Isle Royale in Lake Superior, can test a moose, then gather in a nose-together tail-wagging circle somehow to consult, somehow to decide that this moose is too tough a customer, and so to abandon it, then certainly greater must have been our powers to consult and decide. But primate wit could not have been enough for as vulnerable a predator as were we.

Our intimate knowledge of one another: our size-up that the hunter on our flank, while strong, was also a bit stupid; our acceptance of decision on the part of an alpha leader who knew more than we could ever know about the erratic ways of the wildebeest before us; our admiration for the young fellow just beyond who, though inexperienced, had already exhibited such daring; our slow, silent closure of the hunting ring, deadly perhaps for somebody. In trust we lived; in trust we died. And the social integrity of our little group was our one assurance of survival.

There is an observation on modern life that, while preposterous in its proportions, cannot be neglected: Our juries include eleven members and a foreman. Our traditional army squad includes eleven soldiers and an officer. In the United States we have nine Supreme Court justices. Rare is the government, whatever its proliferation of ministries, in which more than nine, ten, or eleven ministers combine actual power. Rare likewise is the contact sport fielding a team of less than nine or more than eleven. The Soviet Union's Politburo has eleven members. It has even been suggested to me that when Jesus chose his apostles, he chose one too many.

Is this a social propensity in the male inherited from our hunting past? Only rawest speculation could affirm it. Then is it simply a coincidence? The laws of chance must cast doubt. We do not, cannot, and shall never know. The number eleven may be an evolutionary rule in dispositions of the human male, just as the unlucky number thirteen may be a red, flashing, warning light of historic value. All that we may safely adduce is that in male dispositions of trust and mutual understanding, the limited groups of big-brained man remain in the same range of number as in our small-brained ancestors. And discomposing reflections suggest themselves: We may regard the tendency as an inheritance, like the motor pattern of the overhand throw, from the millions of years of our hunting past. Or we may reflect on a more drastic interpretation: Though from the times of little *africanus*, now being found in Ethiopian stream beds over three million years old, the capacity of the brain has trebled, still with all our technical proficiency our social proficiency has remained about the same.

In his *Men in Groups* Tiger has properly emphasized the male bond, the preference of men to be with men which any cocktail

party will demonstrate. He regards it as the spine of human societies, and looks to its evolutionary origins. The book was published in 1969, just too soon to include as significantly as it might the dispositions of the hunting band. While we know that all-male groups are common throughout many species, we know also that they seldom serve a social function or reveal significant organization. It was the evolutionary way of a particular primate who took to the savannas and the hunt—ourselves—that, making use of the tradition of segregated males, welded it into a subgroup so powerful as to remain with us today.

Contemporary industrial managers may thoughtfully reflect, as did McGregor with his Theory Y, on the efficiency, the cooperation, and the emotional satisfactions of the small group of workers. Thus for millions of years were our innate needs for identity, stimulation, and security satisfied. The sociologist, contemplating the depredations of juvenile gangs, may indulge in similar reflections. City planners and architects, facing the urban wilderness, may recall the rule of eleven, whatever its numerical validity, with despair of just possibly a flash of enlightenment. Supervisors of penal or mental institutions may discover in it clues of more immediate value. The educator may sigh: "Very interesting. Just what do I do next?" The hostess may keep it in mind, along with the collateral indication that men may just possibly enjoy men for reasons other than homosexuality.

The small male group was the way of our coming and, despite all expansion of the human neocortex, remains a factor of satisfaction in modern life. Yet we must keep in mind that the hunting band was created and perfected by natural selection on a field of violent action.

5

The success of the hunting adaptation, as Washburn wrote, indeed affected as evolutionary products "our intellect, interests, emotions and basic social life." But we must not forget that the hunting adaptation occurred far longer ago than a few hundred thousand years, and long antedated the formative period in which our brain expanded. Nor may we be allowed to forget, either, that, unlike the gorilla or chimpanzee with their adaptation to the forest and a vegetarian existence, what we were adapting to was the open, hazardous savanna and a life of violent action.

How long ago the adaptation took place is a matter more of curiosity than of final significance. We may be sure that two million years ago the sites at the bottom of Olduvai Gorge were inhabited by successful hunters, and that is long enough in evolutionary terms. But their skill both at hunting and at fashioning stone tools and weapons would indicate an earlier history. Howell's discovery of similar australopithecines in Ethiopia more than a million years earlier seems logical. What stuns the imagination is a discovery of Leakey's at Fort Ternan, in Kenya. It was a being far more primitive, yet exhibiting many of the characteristics of *Australopithecus africanus*, and it has been accepted by all authorities as a hominid whose evolutionary way had already departed from that of the ape. Its age—also accepted by all authorities—is 14,500,000 years.

You and I, who live to seventy years or so, cannot apprehend two million years, let alone almost fifteen million. But the creature was an ancestor, without doubt. Leakey called him *Kenyapithecus*, and refers to him today by that name. Others, however —notably, Elwyn Simons at Yale—have disputed the name and demonstrated that a fossil found in India in the 1930's and called *Ramapithecus* was an approximate contemporary and was in fact the same being. Competent authority agrees with Simons, and so, according to scientific usage, we must refer— though reluctantly—to Leakey's discovery as *Ramapithecus*. The confusion has obscured the importance of what may be the great scientist's most overwhelming find.

Leakey's Fort Ternan hominid proves that fifteen million years ago our way and the ape's had parted. And it suggests, though it does not prove, that even then we were armed hunters.

In 1960, the year before he made his discovery, Leakey showed me photographs of the site with the simple comment, "You never saw so many bones." And the site, limited in size, resembles precisely the living-sites at the bottom of Olduvai Gorge. The next season the Leakeys recovered 1,200 fossils from the assemblage as well as the remains of the hominid. Two years later I had the opportunity to inspect them in Nairobi. The fossils were so well preserved that it was difficult to accept their age. An ashfall had covered them, and it was like something that had happened last Tuesday. There were antelope bones and horn cores in plenty. The descent of ash had come so quickly that there were bones still articulated, some leg bones still joined at the knee; there had been no time to rot apart before consignment to fossilization's deep-freeze. One browsed through an animal Pompeii, and there was the same eerie sense as at Olegorsaillie. By weird magic, one stood in a past beyond comprehension.

The assemblage strikingly resembled those of *habilis* at Olduvai and *africanus* at Makapan. The fractured portions of the hominid jaw were clean. The molars showed almost no trace of vegetarian wear. The canines were no larger than my own.

I discussed the significance of the canine tooth in *African Genesis*. All monkeys and apes then known have formidable fighting canines. If you will feel along your gum above your own eyetooth, you will find a root out of all proportion to the size of the tooth. It is a vestige, a souvenir from the time when we too had fighting teeth. But we could not have survived, lacking such teeth to defend ourselves, had we not become continuously dependent on the weapon in the hand. The reduced canine is a hallmark of true men and all australopithecines, and it was a telling point in Dart's argument that the australopithecines were armed.

Washburn himself, in the classic symposium volume *Behavior and Evolution*, in 1958 demonstrated with a mass of measurements that since in apes and monkeys male canines are far longer than female, their size cannot be related to such functions as ripping up tough foods. He wrote: "Large canines in the male cannot be related to diet, as the females eat the same things, and the males do not provide food. They are related to dominance in, and protection of, the group, both of which functions were replaced in early human societies *by tools.*"

In his reference to tools, Washburn was following anthropology's custom of not saying "weapons" out loud. The argument, indisputable in terms of man, could scarcely be disputed when controversy concerning the australopithecines rose in the 1960's. Even earlier the ecologists G. A. Bartholomew and J. B. Birdsell had in 1953 demonstrated on theoretical grounds that no hominid lacking fighting teeth could have survived without weapons.

And Leakey's Fort Ternan hominid lacked fighting teeth. Could it be possible that the human ancestor fifteen million years ago already went armed? Elwyn Simons, by now with two specimens from India, showed that reduced canines were a species characteristic, and that the Fort Ternan ramapithecine was not a freak. But in the meantime an alternative interpretation of the reduced canine had come along. And since the issue bears such implications concerning the human way, we had best consider it.

Very recently—in fact, in early 1970—two studies of an extinct ape called *Gigantopithecus* have appeared. His remains have been found in China and India, and he was a monstrous creature suitable for primate nightmare. About nine feet tall and weighing in the neighborhood of six hundred pounds, he first appears about six million years ago in the midst of the Pliocene drought when forests had all but shrunk away. *Gigantopithecus* had adapted to life on the grasslands by eating seeds, small roots, stems. One might call it the hard way, supporting such a bulk with such a diet. And the giant, though an ape, lacks fighting teeth. His canines are flat with his molars. He is an exception to primate rule.

The argument has been advanced that ramapithecine reduction of canines may have come about through similar diet. We cannot yet judge, since we know too little about our Miocene ancestors. But the argument offers little but difficulty. Ecologically it seems impossible. *Gigantopithecus* dates from a time when the world-wide Pliocene drought presented crucial problems to all forest creatures. A physically invulnerable monster, he found a way of life in the grasslands. But the little, vulnerable ramapithecines, almost ten million years older, date from the Miocene, as luxurious a period as the world has ever known. There were forests where now are deserts. A colleague of Simons at Yale has described the Indian environment of the time as one of sluggish rivers, tropical forest, with perhaps some tree-dotted savanna. Why would a primate in such a time have turned to eating seeds? And where would he have found them? High-altitude Kenya was an equally rich primate paradise, with more savannas, it is true, on the higher ridges, and more grassland creatures like the antelope, but everywhere abundant forest galleries in the valleys.

I find it quite credible that in the Pliocene drought a primate to survive turned to the hard way of seed-eating; I find it quite incredible in the Miocene. David Pilbeam, another of Simons' colleagues at Yale, writes cautiously concerning the hypothesis: "Although the canines of *Gigantopithecus* were small like those of the Hominidae, morphologically and functionally they were very different. *Gigantopithecus* canines were large grinding teeth, not the chisel-like slicers of Hominidae." The ramapithecine canines were chisel-like.

On May 11, 1968, Louis S. B. Leakey, that Christopher Columbus of anthropology, weighed in with what may be the conclusive evidence, even though in a preliminary report. At Fort Ternan new excavations at the living-site revealed small areas where antelope skulls had been smashed to extract the brains, and limb bones to extract the marrow. Depressed fractures were of a sort that no hyena or other carnivore could have achieved. There was even a lump of lava with battered edges which presumably had been used to do the job. It could be the first known tool.

Since to demonstrate the antiquity of the hunting way we need not go back to the Miocene, let us leave what may become a most horrendous future controversy where it stands today. All those devoted to the primal innocence of man, to his lack of innate aggressiveness, to a denial that violence is a portion of our nature and is solely determined by environmental circumstance, will rise up in favor of eating seeds, a diet perfect for finches. We do not yet know enough, and they could still be right. What must fascinate me is a question more philosophical than scientific, and becomes the ruin of one of my favorite hypotheses.

When Leakey presented his evidence for meat-eating in the Miocene, flushed down the pipeline like one of Brain's bones at Swartkrans went my deprivation hypothesis to become appropriately fossilized. In *African Genesis* I proposed that our ancestors turned to the eating of meat in the depths of the Pliocene drought, when forests had so shrunk that we could no longer compete with the better-adapted forest ape for remaining fruit, and we took to the savannas and the hunt. We were expelled from Eden. It was a beautiful hypothesis, appealing to a dramatist, and it was widely accepted. Leakey ruined me.

Why, in the midst of the earlier lush Miocene with its abundance of forest, fruits, shoots, buds to enchant any arboreal primate, did we turn to the terrestrial life, to the pursuit of meat and the hazards of terrestrial life? My Pliocene hypothesis was an environmentalist's answer, a deprivation conclusion like "poverty causes crime," and, like so many environmentalist answers, may have turned out to be wrong. But what could have inspired us in a time as luscious as a Rubens nude to forsake the comfort and security of arboreal primate life to embrace the dangers of the hunt and the competitions of a more practiced predatory community?

I recall a conversation with Kenneth Oakley in London before *African Genesis* was published and I had yet to place my own bets. We played with the possibility that the human emergence had taken place not in the deprived Pliocene but in the earlier, affluent Miocene. And I was the one to object: Why should we have done it? Oakley, that most elegant of hard-nosed scientists, yet most imaginative, simply shrugged and phrased his answer as a question: "Adventure?"

Why is man man? We have pondered in this chapter certain human propensities which seem to have originated in our long hunting days. We have inspected the all-male hunting band with its habitual exclusion of females, and the mark it may have left on us despite the coming of the big brain. We have considered the hunting territory, anticipating so clearly even the territory of the salesman. We have speculated on the storytelling origins of language as an integrating and educational necessity in small but functionally divided societies. We have considered loyalty, cooperation, mutual trust as the imperative for order if such societies were to survive. And I have purposely postponed till now consideration of that probable consequence which today so absorbs us, our record not just in the morning papers but in all history of a propensity for violence.

If Leakey's Miocene discoveries are correctly interpreted, then we *chose* the violent life; it was not forced upon us. Environmentalism may explain *Gigantopithecus* and the seed-eating way. But why should the ancestral hominid in the richness of Miocene primate times have deserted his arboreal fruits for the hazardous hunt and meat-eating? Every knowledge of innate needs that I possess suggests that the ancestral primate went up the ladder to stimulation, excitement, identity. In a word: adventure.

No one who has had personal experience with monkeys or apes can discount the innate pressure to explore that invests them. Against all reprimand, threats, punishment, they will still explore and gladly take the place to pieces. When the adventurous Miocene primate, for causes perhaps no more definitive than curiosity, launched himself onto the predatory way, he compounded the exploratory legacies of the primate and the violent satisfactions of the predator. If such was the cause, then the human being stands nakedly as an ultimate consequence.

Adventure may or may not have been the motive that began the human way. Whatever it was that set us on our course, still for too many menacing millions of years we found our daily satisfaction in violence. We attacked, or we starved. We dared, or we were selected out. We adapted, anatomically and physiologically, to the hunt. Our muscular buttocks, unlike those of any other primate, provided us with strength to throw, to stab, to crush. Our flattened feet provided speed and endurance in the chase. Glands that once directed the timid primate to flee rearranged their chemistry to direct the hominid to attack. We became creatures adapted in all ways to the excitements of violent action. Until five thousand years ago there was no other way to survive. And if it was only then that organized warfare became a significant human entertainment, perhaps we may understand it as a substitute for the lost hunting way.

I have little patience for those who regard the human propensity for violence as unrelated to our hunting past. If we think of our adaptation as one of appetite and capacity for violent action, then it becomes apparent why man will pursue man as eagerly as he once pursued game; and it becomes equally apparent why, in a mere five thousand years, our appetite has not deserted us. With almost as little patience do I listen to the argument that it is the long-distance weapon, freeing man of the necessity to witness with his own eyes the gruesome consequences of his action, that has made possible man's murder of his own kind. No portion of recorded history instructs us that massacre and brutality must be accomplished at greater than arm's length. Nor does prehistory encourage the view.

J. B. Birdsell once asked Raymond Dart how many of his australopithecines met violent ends, and Dart gloomily answered, "All." He may have exaggerated, but if we consider all the fossils we possess from most ancient times, then a remarkable proportion show evidence of violent death. *Habilis* himself died of a little-publicized fractured skull directly on top of his head. Since he met his end on an open lakeshore, one must assume either that he died of a blow or that the sky fell in on him.

377

At Choukoutien the skulls of forty individuals were found in the caves, with few body bones. Every evidence has suggested that *Homo erectus* practiced head-hunting, and that only the heads were brought home. The skulls had been opened at the base to extract the brains, indicating cannibalism. Among the later Neanderthals, murder—perhaps ritual murder—is common. As striking as any example is that of the oldest known *Homo sapiens*, found near Budapest in 1965. Called Vérteszöllös man, he had a brain as large as our own, he lived 350,000 years ago, and he was found with stone weapons similar to those found in East Africa. He had been killed by one of them.

A propensity for the violent solution is scarcely new in our kind. And if we regard it as just one consequence of our adaptation to violent action demanded by the hunting life, then much becomes clear in our times. Not only murder but riot, assault, vandalism, destruction of property may be seen as violent actions satisfying an appetite without which at one time we could not have survived, but for which little socially acceptable nourishment exists today. And much becomes clearer, too, concerning the attitude we must take toward a force within us that so seeks our destruction.

Should the most hopeful interpretation be correct—that what has come to us through evolutionary legacy is less the need for violent action than the need for the adventure that it satisfies —then we may glimpse our course, whether or not we have the will to take it. Thoughtlessly, unconsciously, yet systematically, modern societies destroy whatever opportunities for adventure may exist. The young mother, bearing a child, faces an overwhelming adventure lacking neither prize nor hazard. Yet the feminist decries motherhood as an unworthy female ambition, devalues the experience. The young man going out into the world faces an intellectual conspiracy advising him that competition is demeaning. And another adventure is devalued. Most smothering of all must be our dedications at the obese altar of the Gross National Product. Just so long as material values surmount all others and material efficiency must be reckoned as our highest morality, then the devouring organization must swallow us. And adventure for the individual must vanish into forgotten scenes of gladiators, clipper ships, Horatio Alger, the North Pole, Polynesian maidens, the frontiersman, Vasco da Gama.

I do not know that our present course is reversible. Neither, however, do I know that our demand for adventure lies at the heart of things. It may be safer to assume that a disposition to take the violent way comes to us just as we see it and history records it, without subtle evolutionary overtones. It is the way we are, and we always shall be. And we must live with it.

If we see, clearly and with conviction, that every human baby born bears the potential resources of the arsonist, the vandal, the murderer, then we shall raise our children differently. If our educational philosophy accepts individual responsibility, not social guilt, as the final determinant of conduct, then we shall see some remarkable changes in the curriculum presented to our students. If we, social members as a whole, agree that no longer shall we applaud the violent, no longer shall we extend our charity to the violator while we ignore the violated, then a quite simple event may take place. Violence, whatever its temptations, could go out of fashion.

The suggestion may seem Utopian. But social attitudes have successfully reduced other social threats. There is no living people lacking a tabu against incest. Yet incest must tempt in every human family. By such a tabu, and by no other means so far as I can see, shall we subdue the violent demon that lies within us all.

As a people normally gets the government it deserves, so a society normally receives the punishments it asks for. And so long as we support the Age of the Alibi, just so long must we inhabit the Age of Anxiety. There must come a limit, of course, when the social order to endure accepts violent means to suppress violent disorder. And we shall then see an endless procession of concentration camps, death penalties, public whippings, and police ascendancy. It is the likelier outcome, no doubt.

379

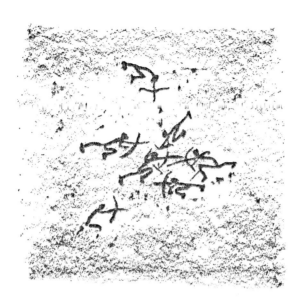

10. The Risen Ape

The years passed by the millions, as indistinguishable one from another as stones on a shingle shore, or scattered, flat-topped thorn trees on a sun-struck, shimmering savanna. Two branches of our early ramapithecine family had emerged. India's sluggish streams and tropical forests vanished. In the high East African wonderland, where green pastures had crowned sinuous hills, and rich forest galleries had crowded the valleys below, the rains came less regularly or not at all. We held on. Through a thousand generations environmental change might have come unnoted, so slow was it. Then, noted or unnoted, some ten or twelve million years ago came the hellish Pliocene. The world dried up and none knows why. But Eden was gone.

Now the years passed by like a procession of mourners in slow single file. The Indian branch of the human experiment failed, became extinct. Perhaps—just perhaps—they had not accepted the meat-eating way and with the vanishment of forest and fruit they expired. We cannot know. Our African line persisted, and this we know or you and I should not be here. We accepted the grassland and its copious population of prey. Inadequate, even ridiculous though we may have seemed in the eyes of the natural predators, still we existed, we survived, we outlasted the drought.

No contemporary mind can visualize such a time of trial. This little group vanished into dust; that little group persevered. What was the distinction? Social order, I suggest. We entered the African Pliocene as ramapithecines, without even, for all we know, satisfactory bipedal carriage. We emerged many million years later as the anatomically developed australopithecine, but with a brain still little larger than the ape's. We were men, up to the neck. But I do not believe that anatomical improvement was the major instrument of our survival; it was rather our social capacity to act as one.

Order far surmounted disorder in the social contract of such times. Sir Julian Huxley once described man as "by far the most variable of wild species known," and I believe that he was correct. The hominid may not, however, have known quite such genetic diversity. Mating must have been confined to a few neighboring societies. Inbreeding of small populations within an unchanging environment over eternally long periods would have tended to reduce variability of individuals. Still, as we know, a random diversity would always have occurred, and it is this diversity that social order would have discouraged. A harsh, predictable environment, demanding today's answer for tomorrow's trial, placed small selective value on the deviant, on the innovator. And society became nature's implement in shaping the genetic diversity of its members to a common sort. The weak, the mentally deficient, were eliminated by infanticide; the hominid could not afford them. The rebellious, if any existed, were driven away to solitude and death. Can one bring to mind today's dissenting youth in a time when the wetness of a waterhole meant the difference between death and life?

A nation of men at war, fighting for survival, will tighten its social contract, renounce individual demands, exist as a single group. Hominid societies throughout the entire Pliocene consisted of tiny biological nations fighting to survive. And the quality of the group, not the distinction of the individual, was the criterion for survival. Cooperation, obedience, dependability, predictability were the qualities of individual merit promoting group existence. Selection, in a word, was for the mediocre.

I find no other persuasive explanation for the failure of the hominid line, through such an expanse of evolutionary time, to do anything much but survive. Our essential demand for social order virtually eliminated that necessary disorder giving room to individual assertion. When the Pliocene at last ended, and the rains of the Pleistocene came, we find *habilis* at the bottom of the Olduvai Gorge. He now had stone tools and weapons, it is true. But 12,500,000 years had elapsed since Fort Ternan, and the way of life seemed not otherwise much different. And now a greater mystery was to appear, a compound mystery of a sort.

Why did the great human brain so suddenly appear, and why did its appearance leave so little mark on our lives? Much later than the original *habilis* in Olduvai's book of pre-history the Leakeys found another example of the being, with a brain a shade *smaller*. It was a million years old. Yet from 350,000 years ago we have the Hungarian with his skull caved in and a brain as large as our own. What happened in little over a half-million years that had not happened before? And why did the appearance of this magnificent new organ present so little change to the human way?

We were now true men. We made somewhat better weapons with a 1,400-cc. brain than had *habilis* with his 650 cc. But his had been adequate. We gained control of fire in Europe and Asia, though not for a long while in Africa. We hunted much the same animals, perhaps killed one another more frequently. As time passes, there is growing evidence for ritual in Neanderthal burials. But if we may guess that the brain had accomplished its expansion by 500,000 years ago, then it was a long time indeed before it produced any miracles.

I should like to present an hypothesis: Not until the invention of the bow and arrow was the individual freed from the social order commanded by the cooperative hunting band. In human history it was the long-distance weapon that made possible the invention of the individual. Our ancient prison of conformity was broken open. The risen ape soared.

The story of the bow and arrow is little known. Perhaps anthropology's insistence on regarding the weapon as just another tool has obscured its significance. One must find it remarkable, for example, that just a few years ago, at Chicago's giant symposium *Man the Hunter*, there was presented no paper, no discussion, not even so far as I can discover a mention of what the coming of the long-distance weapon may have meant to our history.

For what evidence I may present I am indebted to my friend Kenneth Oakley and his irreplaceable works *Man the Tool-Maker* and *Frameworks for Dating Fossil Man*. The bow and arrow was invented in the area of the Atlas Mountains in North Africa at a time when the Sahara was as hospitable to game as is the African savanna today. In this lost green hunting paradise some twenty-five or thirty thousand years ago there still remained a Neanderthal people who created a culture known as the Aterian.

Europe, just across the Mediterranean, was at this time something other than a paradise, since it lay deep-frozen in the grip of Würm, the last great Pleistocene glacier. There Neanderthal had been everywhere replaced by men of our modern sort. Some of us, indeed, were already beginning to scratch and daub on the walls of caves, preparing for the Magdalenian fulfillment of our first true art. And just across the way, in the green Sahara, the last of the Neanderthals was preparing his legacy to the world: death at a distance.

The invention was simple. No one ever before had chipped stone weapons with a base that could be neatly fastened to a shaft. The French, who have given much study to the Aterians, speak of such artifacts as *pièces pédonculées*. In English they are usually called "tanged," and in appearance they resemble the familiar arrowhead of the American Indian. Some were large and, fastened to a shaft, created a throwing spear far more effective than any that could have been made previously. Most were small, however, and could only have been heads for arrows.

We do not know how far these Neanderthal hunters ranged, but Spanish students have found the tanged points in the deposit of Parpallo, near Valencia. Later on in Spain the bow and arrow made its first appearance in rock paintings. By then, overcome by deficiencies that we do not entirely understand, Neanderthal was probably gone, but his legacy remained. Commonly the paintings show men stalking game. But one from a rock-shelter at Castellón shows men wildly fighting.

In Oakley's opinion, the bow and arrow did not spread far or widely replace the spear until after the retreat of the Würm ice sheet. That retreat occurred abruptly, just eleven thousand years ago. In another thousand years we were beginning our domestication of foodstuffs in the Middle East. Modern man was on his way.

The invention of the bow and arrow, I believe, had as much significance to prehistoric man as the invention of the nuclear weapon to modern man. Our relation to the environment was irrevocably altered. We had lived until then, even after the big brain, as one animal species among others. We were superior predators, superior in our wit, in our social capacity, in communication, in the ability to store information through social tradition. But we hunted in a fashion little different from that of the wolf or the lion. Our weapons were simple substitutes for the fangs and the claws of the natural predator. That prey animals should have feared us more than they feared our competitors seems doubtful; that predators feared us at all seems unreasonable. But when the bow and arrow and the far-thrown spear made death at a distance a new fact in the life of the animal, men for the first time stood apart. We became beings of mystery in animal eyes, beings of dread in animal memory. It was a new world.

It was a new world for men, as well. Now the lone hunter could strike down game with small risk to himself. Soon perhaps, like Pygmy or Bushman, he was using arrows tipped with poison. The hunting band was no longer compulsory. No longer had we the desperate need for one another, whether to achieve success in the hunt or safety when we were the hunted. And so today few hunting peoples resort to the old-time band. Their groups are smaller. Colin Turnbull describes the Mbuti Pygmies who live in the Congo's Ituri forest and follow two hunting traditions. There are the archers who with their poisoned arrows hunt in twos or threes and live in small groups. And there are the net hunters who must have many hands to wield the single net, who live communally in larger groups, and who must somehow get along together. Theirs is the old-time way, but it is rare.

With the bow and arrow, I propose, the individual was born as a human possibility. Perhaps the family as we have known it came also into existence, since in the time of the hunting band, if the family existed at all, its significance as a social unit could only have been one of far lesser order. But now one or two men could support wives and children without dependence on the band. And while we still of necessity lived in social groups, an entirely new contract came into being. Natural selection, for so long intolerant of the deviant, could now encourage diversity, search for values other than conformity and mediocrity, and favor social groups which within their contract held a structure of disorder.

I do not know that there is a relation between hunting and the agricultural revolutions. But it would be the new kind of man, capable of innovation, who would have been capable of the farming invention. And it would explain a coincidence which has never been satisfactorily penetrated: why domestication of food supply, after all the long wait, took place independently and quite nearly simultaneously in the Middle East,

Southeast Asia, and the Americas. The bow and arrow, so far as we know, preceded it everywhere.

We need not take speculation so far. What seems evident is that the big brain was of little more value than the small so long as our limited, hand-held weapons perpetuated cooperative hunting. Order was all, and the capacity to conform the supreme selective value. With the invention of the individual and the creation of a new social contract, the brain was released from its social chains. Humanity exploded.

Why the brain itself exploded so many hundreds of thousands of years earlier is of course another question.

2

Why did the brain enlarge? There is a surmise so wild that none with a reputation to lose dares publish it. Yet it is a story so good that to deny it to readers becomes a criminal act. And so, since I lose my reputation anyway as regularly as oak trees lose their leaves, I present it here as the Ardrey Theory of Man the Cosmic Accident. I do so, however, with the strict understanding that I do not believe a word of it.

Seven hundred thousand years ago the earth suffered a violent encounter with a celestial object perhaps a thousand feet in diameter. And this is no invention of mine. The object, probably a small asteroid, entered our atmosphere at an unknown point, but seems to have exploded somewhere west of Australia. Fragments, glassified by the intense heat, are found scattered from Japan to Madagascar. Geologists call them tektites, and the area of their dispersal is about four thousand by six thousand miles. Simultaneously the earth's poles reversed. Before 700,000 years ago a compass would have pointed south. Since then it has pointed north.

Earlier reversals of the earth's magnetic field have occurred at random intervals as far back as geologists can trace. Why they have occurred we do not know; since study and speculation concerning them has absorbed the sciences only in the past few years, it becomes a wonder that we know anything at all. Cores drilled from the sea bottom, however, reveal that during the course of a reversal the earth lacks for about five thousand years any magnetic field at all. And it is that field which provides protection from incoming cosmic rays.

In 1963, an early date in our new studies, the chairman of Canada's defense research board, Robert Uffen, presented the hypothesis that these periods of reversal when life was exposed to cosmic blast could have been times of rapid mutation, appearance of new species, extinction of old. Many such unexplained periods exist in the evolutionary record. No satisfactory explanation has ever been advanced for the sudden die-off of reptiles at the end of the Cretaceous. In the mid-Miocene, coincident with the hominid emergence, there was a sharp, worldwide change of species affecting even such long-time citizens as mollusks and reef corals.

Insufficient time has elapsed for discovery, study and correlation of ancient reversals and times of biological change. Uffen's hypothesis remains a startling idea. But a group of geologists at Columbia University's Lamont Observatory, headed by J. D. Hays and N. D. Opdyke, have taken the idea far beyond the parlors of fancy. Their specialty has been the study of those microscopic creatures known as radiolarians whose mortal remains keep drifting down to the bottom of the sea. From deepsea cores drawn up from the Antarctic bottom they have obtained a biological record of the last five million years. And four distinct faunal zones separated by marked shifts of radiolarian species correlate with magnetic reversals. Since the last reversal, 700,000 years ago, one deals largely with recent species. The 700,000-year marker is distinct.

Opposition to the reversal hypothesis on the part of physicists has been considerable. Absence of the magnetic field would not, in the opinion of some, have brought on a spectacular increase in cosmic rays; in the opinion of others, cosmic rays do not have that significant an influence on rates of mutation. The recent successful adventures of our astronauts beyond the protection of the earth's magnetic field would seem to confirm the physicists, though the subjects were few and they did not stay out there for five thousand years. But the cosmic-ray issue is not critical to the Ardrey Theory of Man the Cosmic Accident.

What happened 700,000 years ago, weirdly recalling Velikovsky's *Worlds in Collision*, was our collision, beyond contradiction, with something big. That a reversal took place simultaneously may have been a coincidence, though the improbability is sky-scraping. Beyond argument is the collision itself, when it took place, and the heat generated.

The best-recorded historical event to offer comparison was a meteorite that landed in Siberia's Tunguska forest one early morning in 1908. It was perhaps the size of a bathroom or two. It too exploded with the friction of our atmosphere, flattening and burning the forest over an area twenty-five miles in diameter, and killing reindeer even at the area's periphery. It has been estimated that the energy released was the equivalent of that which blew up the volcanic island of Krakatoa in 1883. All over the destroyed Siberian forest the same little tektites were found. But Glass and Heezen, of the Lamont group, calculate that our visitor of 700,000 years ago weighed a quarter-billion tons and distributed its fragments, not over a twenty-five mile area of Siberian forest, but over some twenty-five million square miles of the earth's surface.

We regard contemplation of the consequences of a nuclear war as thinking about the unthinkable. Yet as unthinkable was the cataclysm that occurred somewhere over the Indian Ocean. If the energy released by the Siberian disaster may be compared to Krakatoa, then what must have been the energy released by the explosion of the quarter-billion-ton monster? And what must have been the rise in world temperature as a consequence of such trial by fire?

A quarter of a century ago Raymond Cowles investigated the

effect of abnormal temperatures on the male germ cell. Cold temporarily suppresses the formation of sperm, and that is probably why in sudden cold the scrotum shrinks, pressing the testicles within the protection of body heat. But abnormal heat— tested in such a variety of creatures as fruit flies, sparrows, and man—has a twofold effect. It reduces male fecundity, but likewise increases variation. And Haldane, in his *Causes of Evolution*, cites evidence that while heat tends to kill most ova, it induces high mutation rates in the remainder. A proliferation of variants in the male sperm and a plethora of mutation in the female egg: disregarding the effects of radiation, would these have been the consequences of a day of fire 700,000 years ago?

Was the human brain an accident? I have said that I do not believe a word of it, and there are mighty objections. Any such radical transformation of the emerging hominid should have induced radical changes in other animal species. Aside from the Lamont group's record of what happened simultaneously to their Antarctic radiolarians, we have no world-wide observations. The matter has, however, been one of long fascination in Kenya. Suddenly, at approximately the same time, giganticism overwhelmed many animal species. In Nairobi's Coryndon Museum you may look with awe at the fossil remains of giraffe legs that reduce contemporary giraffes to the scale of ostriches. Or you may look at the remains of ostriches that make our contemporary friends look like storks. You may linger before *Bularchus arok*, with a hornspread as wide as a two-lane highway, and compare him with the modern ox. You may gaze upon *Pelorovis oldowayensis* and compare him to the merino ram. His hornspread is quadruple. And he comes from the Olduvai site called BK II, precisely the period of human reference. If you have the time, you may even consider a large tusk of the period, once classified as elephant. It has turned out to be the tusk of an extinct wart hog.

All of this comes from East Africa, the long-standing scene of human evolution. For many years—ever since in 1957 Leakey first showed me his collection of animal grotesques—I have played with the notion that some local circumstance—vulcanism, for example—may have in some wild way contributed to the sudden giganticism of the human brain. But there seemed no adequate explanation, since volcanoes in East Africa have been belching regularly for as long as our ancestors were around. Now today we have an explanation, wild though it may be. We have even East Africa, its nearness to the Indian Ocean and the celestial conflagration, its high altitude which would have afforded minimum atmospheric shield against heat. And while there is the chance that *Homo sapiens* suffered his transformation in Europe or elsewhere, still that oldest of Hungarians, Vérteszöllös man, had weapons of definite African style. So did that oldest of Englishmen, just a shade more recently arrived, Swanscombe man.

It is a problem too complex to settle with evidence available at present. Yet the date of the cosmic cataclysm, falling almost precisely halfway between the last we know of the small-brained hominid and the first we know of big-brained man, must remain, like the ghost within the castle walls, unseen and unprovable, knocking now and then to disturb our sleep. Were we an accident? The knocking carries a peculiar awe if we view the more logical explanations for man.

What might be called the functional explanation for the human brain satisfied me, as it has satisfied most students, until Leakey's demonstration of meat-eating at the ramapithecine Fort Ternan site. Why, for fifteen million years, was the hunting way probable, the demands of cooperative hunting for a better brain logical, and why did nothing happen till a million years ago? And then why did it happen in an evolutionary quick-step? Functional necessity we may grant, a selective advantage for those hunting groups or populations possessing a better brain. Randomness of mutation we may grant, and the necessity to wait until change came our way. But we waited a fearsome long time. And when at last the brain happened, why did it happen with the suddenness of a high wind and an empty garbage can in the middle of the night clattering noisily down a long flight of wooden steps?

I am disenchanted with the functional approach. That we needed better coordination of hand and mind in the making of weapons and tools, that we needed greater areas of memory storage the better to proceed along our hunting way, that, above all, we needed neural centers for communication which the small-brained hominid could not offer—all may be granted. But that little happened when we got the big brain defies functional logic. The humiliating truth must be considered: the arrival of that giant organ to which we attribute human ascendancy had no more effect on our way of life than a fair raise of pay at the office.

If those miraculous mutations which combined to present us with the brain that is ours provided striking selective advan-

tage, then with singular modesty they failed to display the worth
of their wares for a good half-million years. I have presented
my hypothesis that until natural selection could turn to the
individual, then the big brain was a cheap resource. And we
face an evolutionary problem of the cart before the horse; or
the Rolls-Royce before the invention of petrol.

The big brain must remind us of the gift of a shining Rolls-
Royce in the heart of the Pleistocene. We admired it, I am sure.
We enjoyed its shiny look and its deep upholstery. We pushed
it around a bit, marveling that it could move, sat in it, wondered what on earth it was for. Yet not until the invention of
that fuel known as the individual could we with shock recognize what a Rolls-Royce was for.

Why the big brain? It defies all theory of natural selection,
which suggests that those beings of superior endowment will by
immediate demonstration of their superiority survive in greater
number than their inferior predecessors. The Pleistocene exhibits no such record. There is pre-adaptation, the idea that
change may come for which no value will be apparent until a
later date. But for a change so enormous, it is about as unsys-
tematic an evolutionary concept as the Ardrey Theory of Man
the Cosmic Accident. And there is a far more reputable scientific concept of evolutionary advance stemming from the work
of that genius geneticist, Sewall Wright. He saw widely separated interbreeding populations, such as we had in the Pliocene,
as each developing a gene pool of excellent local adaptation.
If with some environmental change these populations, long
separated, came into breeding contact, then genetic explosion
could result. Such was precisely the situation when Pleistocene
rains brought mobility and contact to imprisoned Pliocene human populations. In Wright's terms, anything could happen.

If you want to bet sensible money, bet it on Wright. If you
want to bet crazy money on the field horse, on that bastard heir
of unknown genes, bet it on the Ardrey Theory of Man the
Cosmic Accident. I say I do not believe it. Yet it demonstrates
just about as much sense as do we.

3

Arthur Koestler has gloomily proposed that there must be something wrong with the human brain. Hastily I rise to second the motion, as must virtually any audience that reads the morning papers. The risen ape too frequently shows signs of confusion as to which way he is headed, up or down. As Konrad Lorenz once commented, *Homo sapiens* still remains the halfway house between the ape and the human being. But I am not quite so gloomy as Koestler. Perhaps this is because I am not a Hungarian with an ancestor who, after half a million years of hoping for the best, must by now be a most disappointed fossil. I am instead a *nouveau* barbarian sprung from Scottish ancestors who until a century and a half ago delighted in nothing so much as killing one another and in this short lapse of time have at least made a certain civilized advance. If what has happened to the Scots can happen in New Guinea, then even Papuans have hope.

Koestler too, with his acute perceptions, has seen that, whatever the evolutionary causes, man received his brain too fast. "Evolution proceeds by trial and error, so we ought not be surprised if it turned out that there is some construction fault in the circuitry that we carry inside our skulls that would explain the unholy mess we have made of our history." An error in circuitry should not be surprising. How many millions of years did it take us to perfect the erect posture? Yet the imperfect sacroiliac remains for many a human curse. For how many millions of years have we scampered about on two feet? Yet fallen arches remain a most immediate human temptation; I suspect my own of falling in dismay with my first step. We are evolutionary experiments, and the wonder to me is not that we have done so badly but that we have done so well.

It has been the work of Paul MacLean that has advanced more than that of any other specialist our understanding of just what went wrong in the brain explosion of the mid-Pleistocene. MacLean is a neurophysiologist, and, like John Calhoun, is with our National Institute of Mental Health, where he is director of limbic research. "Limbic" refers not to the neocortex, the scene within our skulls where the explosion took place, but to the more ancient sub-brain in which not too much happened. Perhaps a fundamental error is our tendency to regard the brain as a single organ like the heart or liver. MacLean takes the evolutionary view that the human brain has gone through three stages of advance which give us what amounts to three brains.

The oldest and most central portion is what he calls the reptilian brain, containing the brainstem and certain basic ganglia. It is this brain that once programmed certain settled ways of meeting situations. "In other words," writes MacLean, "it seems to play a primary role in instinctually determined functions such as establishing territory, finding shelter, hunting, homing, mating, breeding, forming social hierarchies, selecting leaders, and the like." Precedent is the guiding force. "It would be satisfying to know to what extent the reptilian counterpart of man's brain determines his obeisance to precedent in ceremonial rituals, legal actions, political persuasions and religious convictions."

It is our failure to understand that the human brain is not all lively, rational cortex that makes it possible for many to reject the animal within us. Evolution proceeded beyond the time of the reptile, for the defect of the oldest brain was its clumsiness when confronted with novel situations. But frugal nature, as MacLean says, threw nothing away. The reptilian brain remained within us, buried by new cortical accommodations, yet still retaining, like a storeroom full of memory's old gear, patterns ready at a moment's notice to enforce or disrupt human behavior.

The evolving mammal brought to the natural world not just babies, fur, a central heating system, and new dental arrangements, but what MacLean calls a "thinking cap" as well. It was a lobe of primitive cortex surrounding in a ring the original reptilian brain. It is found in all mammals. Perhaps a hundred million years ancient in its origins, it had ample evolutionary time to perfect its connections with older reptilian installations. The two animal brains, well integrated, constitute what in man is called the limbic system.

The addition of the primitive mammalian cortex meant keener capacities to learn, to adapt old ways to new environmental challenges, to feel and express wider and more sensitive ranges of emotion. The sense of smell, so important to all mammals other than the advanced primates, became intimately integrated with sexual activities, identification, various acts of self-preservation ranging from fear to feeding. The hypothalamus, a reptilian legacy serving as a kind of emotional switchboard and mediator, developed neural connections with the limbic cortex so strong that some nerve bundles are as thick as a lead pencil. No such striking connections exist with the neocortex. Some pathways are so fine that the nerves have never been traced.

We have not had time. The mammal has had a hundred million years or more to perfect by evolutionary trial and error the integration of the two old animal brains. The neocortex, the third of the animal brains, appeared as a distinctive feature of monkeys and apes, and this was what was slowly expanding until the time of *habilis* and the other australopithecines. And it was in this new structure, not the old limbic system, that the human explosion took place. We have had a mere half-million years to perfect its connections with our cerebral inheritance from the animal.

As if the neurophysiological problem were not enough, MacLean points to another. The new brain speaks in a language that the old brain does not understand. With the gargantuan neuronal resources of the human neocortex, extensions of foresight and memory, symbolic language, conceptual thought and self-awareness become possible. But the animal brain does not know the language. Through moods and emotions the old brain can communicate with the new. But only with the greatest difficulty can we talk back, for it is precisely the equivalent of talking to animals. And that is why, for example, we may understand perfectly the cause of a psychosomatic affliction and just as perfectly be unable to do anything about it. We "act against our best judgment," "let our worst impulses get the better of us," plead that "somehow or other we could not control ourselves."

It has been the surgery of Dr. Paul MacLean that has laid open the ancient animal still investing us. And like all good surgery, having exposed the defect, it prepares the cure. Animals, as we all know, can at least to a degree be tamed.

For those who persist in denying the evolutionary influences on human behavior, there is truly little hope. The animal within us, whose existence is denied, whose ways are ignored, or whose presence if suspected is secretly hated or feared, remains

a wild animal. But the animal who is accepted, whose ways become known to us, to whom we speak in his language rather than ours, may become a tame animal. So Hediger's lion-tamer, with most intimate understanding of lion ways, exerts control over his formidable companion.

We can no more deny the animal within us than we can deny the wolf in the fold. Yet the wolf is tamable, if not too trustworthy. And granted a few thousands of generations of affectionate, understanding relationship—who knows?—the wolf may become a dog.

It is truly a hope of Utopian dimension, and in any case offers small comfort for man and his anxieties, today and tomorrow. We remain the wild species that Huxley described. What understanding and taming of the animal within us we can achieve is an acquired characteristic, and cannot be transmitted to our descendants. With every generation born, we must begin anew. But if we deny the wolf, then we shall have nowhere even to begin.

4

As we review the effects of twentieth-century evolutionary thought on our earlier and still cherished social philosophies, I find one upset the most remarkable. In the eighteenth century we could not conceive of social orders prevailing in nature. Whether we took the view of Rousseau—that of original man strolling alone and at peace through the forest—or the still earlier view of Hobbes—that in primal times it was everyman against everyman—we conceived of the individual as the ancient reality, and society as the human invention. Yet the broadest and most indisputable conclusion must be that society, for almost all, and for always, has been nature's cradle. Social order —with its rules and regulations, its alphas and omegas, its territories and its hierarchies, its competitions and xenophobias —has been the evolutionary way. And if I am correct, then it is the individual as we know him that has been the human invention.

Could we a mere fifteen thousand years ago have described man as the risen ape? Well, yes, for there were those grand achievements of painting being executed at that very time on the walls of such caves as Lascaux. And while the chimpanzee's talents are of no mean order, as Desmond Morris has shown us, still no chimp could have given us the bulls of Lascaux. And there were those Neanderthals vanishing forever in the green Sahara but leaving us their legacy of death at a distance. Assassins would become possible, and Shakespeares. But one cannot say that much had happened yet. The ape had risen, but not very far.

Man indeed was born yesterday. Social order is contained in our animal past and is ingrained in the patterns of our animal sub-brains. The individual is the creature of the human future, and we still do not know quite what to do about him. Arthur Koestler, with another of his flashing insights, has written:

> What I am trying to suggest is that the aggressive, self-assertive tendencies in the emotional life of the individual are less dangerous to the species than his transcending or integrative tendencies. Most civilizations have been quite successful in taming individual aggressiveness and teaching the young how to sublimate their self-assertive impulses. But we have tragically failed to achieve a similar sublimation and canalization of the self-transcending emotions. The number of victims of individual crimes committed in any period of history is insignificant compared to the masses cheerfully sacrificed *ad majorem gloriam*, in blind devotion to the true religion, dynasty, or political system.

It is a statement of dazzling heresy, for Koestler is denying ultimate evil as rising from the greed and selfishness of the individual, and looking to those qualities which we regard as self-sacrificing, dedicated, rooted in social action, as the consuming forces that we have never controlled. The heresy must be inspected with the same thoughtfulness that propounded it. For if the social mechanisms are a portion of our animal legacy, then we may understand why they are so difficult to control. And if the individual is malleable, it is because he is a human invention. When we speak to him in the language of the neocortex, it is a language he understands.

We may think of the mob. What incites it but animal language? Shouts, rhythms, loaded words, gestures to rival the upraised tail, the hot symbol of a cross, a swastika, a dummy hanged in effigy. A mob transcends its leaders, becomes a single wild happy thing satisfying identity, stimulation, the following response, xenophobia, australopithecine joys of the hunt and the kill, a thing that through delirious social self-approval discards all neocortical inhibition. To describe a mob as subhuman is incorrect; it simply ceases to behave like individuals. To regard it as a storm of disorder is equally incorrect, for a mob is as orderly a human phenomenon as one will ever encounter: let a single voice of rational dissent be raised within it, and observe what happens to the dissenter.

The mob may with all propriety be described as a monster, since whether its object is to lynch a Negro, destroy a college building, or in the name of Jesus Christ to kill other Christians, its essence is reptilian. And while Koestler recalls the social crimes of human history, whereas in this inquiry my attention is directed to the civil violence of present and future history, still we describe the same phenomenon: those obscenities uncontrollable by individual reason released by the sub-brains' ancient resources of social response.

We tamper with the individual at utmost peril, for it is he, not the mob, who can learn. It is he, the post-neocortical inhabitant of our skulls, who possesses the foresight to make alliances when alliances are to his interest, to make compromises when compromises offer practical advantage, to inhibit the violent action when violence in the end will only destroy him—who, though tempted to assault society, will still ask how he will survive without it. The reptilian mob possesses no such capacity.

The individual may and must grapple with the animal within him. As Anthony Storr has written, "Although we may recoil in horror when we read in newspaper or history book of the atrocities committed by man upon man, we know in our hearts that each one of us harbors within him those same savage impulses which lead to murder, to torture and war." The individual with his post-neocortical self-awareness can know it, accept it, guard against it. The reptilian mob cannot.

The individual, with his post-neocortical creation, speaks the languages of men. He may not accept reason, but he can listen to it. A word may carry a different concept for you than it does for me. Still we may debate it. I may speak Zulu, you Swedish, but we may hire an interpreter. And our logical processes will prove not that different. Parent and child, speaking from worlds of experience as far apart as stars in a galaxy, may still sit down together and discover solutions—if the mob has not intruded. But the mob speaks no human language.

We repress the individual at our peril. We erect monolithic states in gigantic imitation of the hominid hunting band, direct selection, as in the ancient past, to the survival of the mediocre, but we do not have evolutionary time to make such selection a working proposition. And so no such state exists without police power to enforce conformity on the individual. Or we encourage organization, diminishing and ever diminishing the roles that individuals may play, the sovereignty that individuals may strive for, the dignity and the independence and the confidence that a man once felt as he grasped his bow and arrow.

It was the individual who created our civilizations. After millions of years of social repression the individual, released, released the great brain. There was an Egyptian long ago named Imhotep, and he was the world's first architect. He built our earliest remaining masonry construction, the step pyramid at Saqqara, and five thousand years later it yet dominates the sands. Individuals have given us not just pyramids but poems, philosophies, rebellions, the reading of stars and atoms, villains, heroes, vendors of death, vendors of dreams, Iagos, Othellos, the nostalgic remembrances of a Proust, the dubious anticipations of a Plato, the confirmations of a Julius Caesar, the dark ambitions of a Hitler, the black doubts of a Dostoievsky, the high pragmatism of a Lincoln, the corrosive faith of a St. Paul. It is the individual who has brought us that dynamic mixed bag called human civilization. And how high will General Motors stand above the American sands five thousand years from now?

We shall make compromises. Society will recognize, as the social contract dictates, that the individual is the one and only source of human fulfillment. As government is the servant of the people, the organization will recognize itself as the servant of the individual, without whose genius organization would be a fishnet in interstellar space. We shall make elbow room, whatever the price, for individuality.

But the individual will make compromises, too, for he will renounce the mob. And he will grant that men are created unequal.

5

That most unlikely of citizens, George Bernard Shaw, wrote in his preface to *Androcles and the Lion:* "Government is impossible without a religion." He referred to a common body of assumptions accepted by all, and as he referred to government, we may speak of its master, society. There exists no natural society of animals that does not accept certain rules and regulations, certain actions accepted by all that are done or not done. There exists no waterbuck lacking a territory who expects female responsiveness; and so he shrugs off sexual preoccupation. There exists no African buffalo lacking alpha status who presumes sexual attention on the part of the female; he accepts the law of his species. There exists no alpha baboon who, when the

troop is challenged by the predator, will not accept the risk and like a medieval knight go forth to do battle. It is the way things are. There exists no male robin who will not work his wings off to feed his young; no Uganda kob who will not accept that the female on the next fellow's territory is the next fellow's; no African hunting dog that will not honor the obligation of regurgitating a portion of his splendid meal for those, male or female, who stayed home to guard the pups. There is no animal society without a religion, a set of assumptions unquestioned and accepted by all.

For reasons which I regard as comprehensible, we lack such religions. Perhaps since the time of Neanderthal, when we painted our corpses with red ocher, we accepted death as the ultimate union of man. In hope and in fear we created ritual as a reminder of social union. Throughout times of sorcery, oracles, priests, we accomplished social integrity through increasing fear, whether of witches, curses, the Devil himself, or, in later dispensations, faith in a personal God. Such devotions could only be temporary. The human neocortex with its powers of awareness and foresight, implemented as best as possible by the individual, could not but recognize that the personal God I prayed to must be listening to others as well. Arab astronomers, in the time of Islam's ascendancy, must have privately wondered just how much time Allah had to devote to us, with so much else on His hands. From early days, within whatever parochial confines of divine dedication, there must always have lurked those dissenting, wandering merchants or star-bemused hermits. Then the rational inroads of the eighteenth century and the scientific uproar of the nineteenth left visions of heaven and hell, gods and demons as inadequate beings to unite us in faith or fear.

There had been the problem always that our gods had been geographical, territorial, tribal, intolerant of others. Malachi Martin in *The Encounter* brilliantly analyzes the crises of the great religions in their projection of dominance as a character of gods satisfying only their true believers. Never have we known a "religion" to approximate the religion of animals, a set of assumptions accepted without question by an entire species.

There were those of us, of course, who did our almighty best. From the eighteenth century onward we sought a neocortical, man-made faith. Thomas Jefferson's Declaration of Independence proclaimed politically that men are created equal. As a rebel cry to unite the most moronic and most gifted of American colonials in revolution against their masters, it made a propaganda point. Whether anyone ever believed it—and with doubt one must meditate on its sophisticated author—I cannot say. But the natural, innate equality of men became in time the nearest approach to a universal religion that we have ever known. There was a drawback, of course. No one boasting even a minimum of human experience could accept in his heart a doctrine so ethereal except a mass of unequals or a handful of non-terrestrial intellectuals who would quickly shed their dedication if, by chance, power came their way.

Still we did our best. Nineteenth-century man, getting rich himself, compromised natural equality with materialism. When we all had gained enough, then innate equality would become apparent. A religion of material want was something less than the life of the spirit, but so long as virtually everyone was deprived, who could prove that it was not so? And in the meantime the rich got richer relieving the deprived of their deprivations. Had they not been so successful in at least a few societies, we should not today face a brand-new crisis. As peace catches up with war, so affluence catches up with want. And we find hatreds multiplied, inequalities laid more bare.

We are a species lacking a religion. We are members of societies lacking common bodies of assumption. We are philosophical bankrupts. We are ravaged by every mob that moves, every bandit that confronts our way. Guilt rapes us. What did we do that was wrong? For it must be our fault. Anxieties shred our concentrations. How will it be tomorrow? No philosophy, like the Dipper's great bowl, points to a luminous point in space and says this way is north. We are temporal beings in an overbearing universe. The more we know, the more painfully conscious we become of our personal insignificance, of our personal helplessness, of our welcome mortality. We turn to science, which as a species we have come to accept as the contemporary temple. And the priests disagree.

Yet science without question is the universally accepted temple. Science, for the immense majority of the human species, represents our Delphic shrine. We worship it. We consult its oracles. We presume miracles. When men walk on the moon, we bow. Whether or not the priests agree, *Homo sapiens* as a species turns to a single temple, and presumes answers which he does not receive.

A temple exists, acceptable to the broad majority of men and nations, commanding authority once monopolized by gods. Here is the meeting place from which a body of common assumptions might emerge. But too much cannot be expected of the scientist himself, for he is a specialist. If his concern is with the molecule, then his authority diminishes as he approaches the cell. If his concern is with the cell, then his authority diminishes as he approaches the body. If his concern is with the body, then he may be as confused as are we when he approaches mankind. And even though his concern may be with society, his ignorance may be quite impeccable concerning the cell.

What we lack is an evolutionary philosophy. For too long the philosopher has been the uninvited guest at our table. For too long in contemporary life the philosopher has remained a weird and somewhat embarrassing eccentric to whom we give Christmas baskets. And if we are a people lacking a philosophy, this must be a reason why. As our knowledge grows, so does our understanding diminish. We may have those among us who have mastered knowledge of the double helix. We have among us few who have arrived at an evolutionary understanding of man's dual nature. The specialist—whether in the manufacture of motor cars or the manufacture of enzymes, whether in the probing of cosmic origins or in the probing of neuroses—has reacted to the philosopher as any territorial proprietor must react to an intruder.

I am an observer of the sciences, and I cannot speak for the temple. But I may suggest that there is a union of the Visible and the Invisible apparent in the evolutionary nature of man. There is that visible being, the man who sits down before you in need of a haircut, suffering at the moment perhaps from too many drinks last night, brilliant, ambitious, guiltily conscious that his ambitious preoccupations are providing his wife with a somewhat deprived sex life, yet bewildered that he is so attracted by the secretary who, like a female baboon, keeps presenting; unsure that his ambitions will come to anything, yet determined that they will; continually wishing that he could live in the fragrant countryside where he grew up instead of in the disinfected city where he must live; it is a brief portrait of a man.

Yet what you are in the presence of is the geneticist's phenotype, the being with a genetic endowment who materializes before you with adaptations or maladaptations to his environment. He is not a static creature. He is a being with continual dynamic response to the foundations of his genetical endowment and the opportunities or hazards of his environment. He is the Visible. He is what you are having a drink with.

The evolutionist, like a drunk with eyes doubly focused, must see everyone in terms of the Visible and the Invisible. There is the man before you, the Visible, the phenotype. He is the substance. But there is the shadow beyond him, the union of egg and sperm, the accident of the night with its genetic proposal which has been expanded or forfeited perhaps just as accidentally in a lifetime of encounters with environment. It is a harsh sort of inspection that springs from the evolutionist's double vision.

Yet for all its harshness, a philosophy founded on science's demonstrables brings us around to the simplicities of all religions, human or animal. We see the Visible, but we may contemplate the Invisible, and the Invisible differs little from what we once called the soul. Through its unequal consequences the accident of the night divides us. Yet it unites us. We preserve all history, as randomly we dispense the future. We are born, we die as vulnerable individuals, but we carry within us that genetic union, our participation in a population's gene pool. The Invisible is our community and our eternity of interest. And as genetic endowment divides population from population, race from race, yet we need not look back too far in the history of our species to find that time when populations and races were one.

The Visible exists in three dimensions; the Invisible in four. Hatred at its worst is a repulsion between Visibles. Love at its most consummate is a union of Invisibles. Mutual derogation is an acceptance of the instant as all. Mutual respect is an acknowledgment of history. That you and I are here at all is a testament to that dimension which I call the Invisible. That you and I may surrender to hubris represents, perhaps, a contemporary sorting on the part of natural selection among unequals: those who can accept the four-dimensional nature of being and who will therefore survive; and those who cannot, and who will ultimately perish.

The evolutionary nature of man represents, as I see it, a subject for the new philosopher, the new theologian. A set of common assumptions, common dedications, common assurances, of rules and regulations, even considering the limitations of *Homo sapiens*, remains someday possible. As all of our parochial dedications have been eroded by the wash of the sciences, still a religion unassailable by the sciences exists as a goal worthy of contemporary ambition.

But we cannot lie to ourselves. It is the rule of the animal as it must be of men. We shall have our arguments through whatever eternity selection allots us. We shall conduct arrogant experiments with the social contract until biological command interferes. We shall indulge ourselves in such exhibitions of hubris as the intellectual or the mob may designate. In the end we shall submit. We shall accept the only evolutionary conciliation, the philosophy of the possible.

The life that we know, if we listen closely to its music, announces the evolutionary experience. Natural laws unknown to a Jean-Jacques Rousseau, despite his inquiring genius, enclose and yet enforce the human adventure. Natural laws—the shape of the inarguable—while subject today to scientific dispute and denial, must project with sufficient investigation an affirmation of the eternally Invisible, the shadings of time beyond recollection, the definitions of experience that neither you nor I may recollect, the gathered wisdom distilled from the affluent Miocene, compacted by the deprived Pliocene, challenged, frustrated, tempted, rejected by that harlot, the frivolous Pleistocene, until man came forth.

The rhythm of drums from across ancient lakes inspires our

bodies. The chorales of ancient voices, swelling in mystery from beyond geological hills, invest our social being. Yet it is the trumpet—the clean clear clarion announcement of individual entrance and ascendancy—that proclaims the human being as we know him. Our heartbeat rises. The Invisible invests the Visible. We respond, and it is all we know. The bugle crashingly announces the morning; plaintively suggests that day is done; or tragically places its signature, with long withdrawal, on the document testifying to the death of a man.

Dreary will be the morning when you and I awake and leopards are gone; when starlings in hordes no longer chatter in the plane trees gossiping about the adventures of the day to come; when the lone tomcat fails to return from his night's excesses; when robins cease to cry out their belligerent challenges to the bushes beyond the lawn; when the skies lack larks and the shrubbery lacks sex-obsessed rabbits hopping after each other; when hawks cease their eternal, circling searching and the gullery by the rocks falls silent; when the diversity of species no longer illuminates the morning hour and the diversity of men has vanished like the last dawn-afflicted star: if this be the morning we must waken to, then may I, please God, have died in my sleep.

Yet it is the morning that, knowing or unknowing, we strive for: you, I, capitalists, socialists, yellow, white, brown. It is the morning that professors demand in common with policemen, that the philosophies of two centuries have praised, the morning of identicality, of the commonly induced conditioned reflex, the morning of egalitarian actuality, of the brave new world, of order beyond argument, of gray shadows beyond distinction, of uniform response to uniform stimulus, the morning of a tinkling bell and sheep proceeding to pasture. Let me never wake.

It is the morning we praise and we pray for in our industrial organizations, on our collective farms, in our churchly councils, in our processes of government, in our relations between states, in our righteous demands for world government, in our most seemly prayers that someday we shall all be the same. It is the morning that the young, whether they know it or not, rise against in protest. And it is a morning, may the skies of our origins be worshipped beyond measure, that will never come.

As life is larger than man, so is life wiser than are we. As evolution has made us possible, so will evolution sit in final judgment. As natural selection declared us in, so natural selection, should our hubris overcome us, will declare us out. But the stark gray morning will never come to be, for laws larger than you or me will, with impartial, imperishable accord, at some night-court in the course of man's darkness, condemn us as a species to extinction—or more probably will enforce on us the laws of all flesh.

BIBLIOGRAPHICAL KEY, BIBLIOGRAPHY AND INDEX

Bibliographical Key and Bibliography

THE NUMBERED REFERENCES are those to which the BIB-LIOGRAPHICAL KEY is a guide. If the reference is a book, the title is italicized. If the reference is a paper appearing in a professional journal, the name of the journal is italicized, and volume, page, and date are indicated. If the reference is a paper appearing in a book edited by another author, then the editor's name may be indicated, with the reference number in parentheses.

Bibliographical Key

1. TUSKLESS IN PARADISE   pp. 11–37

SECTION 1: pp. 11–12. None.

SECTION 2: pp. 12–18. Amoeba 199.

SECTION 3: pp. 18–30. Orchids 115, 313. Ceylon viper, Madagascar *Langaha, Oxybelis* 123. *Chactodon* 293. British plane 56. Nile catfish 115. Ceylon shrike 236. Watson 308. Skinner on Lorenz 271. Montagu 210. Harlow on learning 116. ULCA 29, 145. Hirsch 126. Skinner's rats 272, domestication quote 271. Definition domestication 27. Sahlins 256. Washburn 304. Lemur 235. Gorilla 257. Lancaster 169. Harlow on sex 119. Chomsky 49. Lenneberg 182. Buss PC. Schaller on elephant 258. Adamson PC.

SECTION 4: pp. 30–35. Darwin 66. Tax 285. Steward 278.

SECTION 5: pp. 35–37. None.

2. THE ACCIDENT OF THE NIGHT   pp. 39–79

SECTION 1: pp. 39–46. General discussion 202, 134, 18, 205, 270. Earliest life 13. Wright on clones 319. Wallace quote and Linnean meeting 96. Malthus 192. Huxley quotes 134. Fisher 93. Haldane 107. Wright 318. Dobzhansky on genes 76. Whyte 311. Sibling chances 126.

SECTION 2: pp. 46–58. Moe 208. Lamarck and Lysenko 44. Lorenz 190. Storr quotes 279. Keith 147. Simpson 270. Baerends in 292. Tinbergen 292. Von Frisch in 289. New Jersey ant 316. Ruff 127. Japanese monkeys 136. Carpenter 40. Koford 153.

SECTION 3: pp. 58–69. Beadle 18. Mayr 205. Dobzhansky 75. Bear macaque 94. Weaverbird 57. Snails 134. Wright in 207, 205. Simpson 269. Calhoun 37. Dobzhansky on isolates 76. Sage grouse 262. Kob

5. TIME AND THE YOUNG BABOON pp. 159–197

SECTION 1: pp. 159–170. Rhesus experiment 255. Beaver 25. Small rodents 36. Schaller on lion PC. Psychological castration 106. Wildebeest 89. Crook-Gartlan 59. Hunting dog 161 and Schaller PC.

SECTION 2: pp. 170–179. Lemur 235. Groos in 155. Durkheim in 54. Rensch 241. Washburn 305. Carpenter 40. Burghers 35. Collias 53. Altmann 8. Piaget in 54. Langur 245. Baboon 304. Rhesus rank 144.

SECTION 3: pp. 179–190. James 139, PC. Hawthorne experiment in 312. McGregor 218. Kuriloff 163, 164. Maslow on motivation 196, on synergy 197.

SECTION 4: pp. 190–197. Baboon 304, 72, 113, 110.

6. DEATH BY STRESS pp. 199–240

SECTION 1: pp. 199–201. Guppies 30. Malthus 192.

SECTION 2: pp. 202–209. Howard 128. Read 240. Dart 64. Carr-Saunders 41. Elton 82. Lemming history 26. Homing pigeons 204. Swedish lemmings 69. Seton in 273. Canadian forests 31. MacLulich 222. O. Kalela in 15. Helsingfors study 233. Chitty 48. California vole 158. Errington 83, 84.

SECTION 3: pp. 209–218. Calhoun 37. Moffat 209. Hinde 124. Red grouse 307. Australian magpie 42. Howard 128. Uganda kob 34. Antelope review 89. Lion 259 and Schaller PC. Elephant 171. Deer 23.

SECTION 4: pp. 219–226. Lack 166. Wynne-Edwards 320. Lack v. W-E 167. Woodpecker in 165. Great tit 168. Extension great-tit study 234. Wilson on competition 315. Kittiwake 60. Gannet photograph 321. Mayr 206. Solomon 273.

SECTION 5: pp. 226–232. Green in 23. Christian in 48. Christian and Davis on physiology 50. Manitoba shrew 33. Bruce effect 47, in laboratory shrew 51. Cockroach 91. Glasgow rat 15. Female rabbits 215. Telephone company 125.

SECTION 6: pp. 232–240. Herring gull 291. Birdsell 19. Carr-Saunders 41. Kenyatta 148.

7. SPACE AND THE CITIZEN pp. 241–283

SECTION 1: pp. 241–247. Ten-spined stickleback 213. Behavioral sink 38.

SECTION 2: pp. 247–254. Russell 254. Langurs: Jay 140, Ripley 245,

Sugiyama 282. Hall's Bloemfontein baboons PC. Vervets: Lolui Island 114, Amboseli 280, 281.

SECTION 3: pp. 254–263. McBride 217. Sommer 274. Taureg 214. Altman 7. Esser 85, 86. Marais 193. Hediger 123. Kinzel 151.

SECTION 4: pp. 263–270. Hediger 123. Rhesus 276. E. T. Hall 109. Suttles 283.

SECTION 5: pp. 270–280. Davis 68. Leyhausen 184, 185. Hall 109. Rabbit 216. Intelligence distribution 103, 322.

SECTION 6: pp. 280–283. None.

8. THE VIOLENT WAY pp. 285–332

SECTION 1: pp. 285–287. *Guardian* editorial 104. White 310.

SECTION 2: pp. 287–299. Berkowitz. Storr in 279. Carrighar 43. Crook 58. Lorenz 190. Gorer 102. Hopcraft PC. Herring gull 291. Skua 81. Eibl-Eibesfeldt 80. Coral snake 191. Isle Royale wolves 142. Lions: Serengeti 259, others PC.

SECTION 3: pp. 300–311. Valley quail 129. Howler 40. Vervet 281. Herring gull 291. Proboscis monkey 149. Langur 140. Buffalo (J. Grimsdell) in 186. Reynolds 242. Chimps: Goodall 100, 101, Kortlandt 156, Reynolds 243. Chimp meat-eating 99. Washburn 306.

SECTION 4: pp. 311–323. Tiger 290. Friedan 95. *Guardian* 104. Suttles 283.

SECTION 5: pp. 323–332. Morris 212. Bushman 265.

9. THE LIONS OF GORONGOSA pp. 333–379

SECTION 1: pp. 333–337. Laughlin 170. Washburn 306. Dart 64. Romer 249. Howell 131, date 230.

SECTION 2: pp. 337 349. Russell 254. Lee 179. Hazda 317. Vallois 298. Fire 230, 231, 287. Hunting and brain size 229, 230, 231. Olorgesaillie 137. *Zinjanthropus* classification 39. Leakey on *habilis* 172. Controversy 181, 230, 39, 130. Olduvai industry 178. Leakey on fauna 173, 174, quote 173.

SECTION 3: pp. 349–358. Jung 143. Adamson PC. Kruuk 159, 160, PC. Schaller on hunting v. scavenging 260, De Vore-Washburn 73. Brain 28. Hominid relevance prey size 260.

SECTION 4: pp. 359–370. Girlish throw 157. *Habilis* footbones 224. Lion cooperative hunting Schaller PC. Stone ovals Olduvai 225, Nice 70.

417

Bibliography

1. ADLER, ALFRED. *Problems of Neurosis.* London: Kegan Paul, 1929.
2. ADLER, ALFRED. *What Life Should Mean to You.* London: George Allen & Unwin, 1932.
3. ALLEE, W. C. Analytical studies of group behavior in birds. *Wilson Bull.* 48:145–51, 1936.
4. ALLEE, W. C. Population problems in protozoa. *Amer. Naturalist* 75:473–87, 1941.
5. ALLEE, W. C. *Social Life of Animals.* 1938. Revised edition, Boston: Beacon Press, 1958.
6. ALLEN, DURWARD L., and L. D. MECH. Wolves vs. moose on Isle Royale. *National Geographic* Feb. 1963.
7. ALTMAN, IRWIN, and W. HAYTHORN. Ecology of isolated groups. *Behavioral Science* 12:169–82, 1967.
8. ALTMANN, S. A., ed. *Social Communication Among Primates.* Univ. of Chicago Press, 1967.
9. ALTMANN, S. A. Structure of primate social communication. In Altmann (8), 1967.
10. ALTMANN, S. A. Field study of the sociobiology of rhesus monkeys. *Ann. N.Y. Acad. Science* 102:338–435, 1962.
11. ALTUS, W. D. Birth order and its sequalae. *Science* 151:44–9, 1966.
12. BANKS, EDWIN M. Social organization in red jungle fowl hens. *Ecology* 37:239–48, 1956.
13. BARGHOORN, E. S., and J. W. SCHOPF. Micro-organisms 3 billion years old from the Pre-Cambrian of South Africa. *Science* 152: 758–62, 1966.
14. BARNETT, S. A., ed. *A Century of Darwin.* London: Heinemann, 1958.
15. BARNETT, S. A. Social stress. *Viewpoints in Biology* 3:170–218, 1964.
16. BARTHOLOMEW, G. A., and J. B. BIRDSELL. Ecology and the proto-hominids. *Amer. Anthropologist* Oct. 1953.
17. BARTHOLOMEW, G. A., and PAUL G. HOEL. Reproductive behavior of the Alaska fur seal. *Jour. Mammalogy* 34:417–36, 1953.

18. BEADLE, GEORGE and MURIEL. *The Language of Life*. New York: Doubleday, 1966.

19. BIRDSELL, J. B. Some predictions for the Pleistocene based on equilibrium systems among recent hunter-gatherers. In Lee and De Vore (180), 1968. Also Discussion.

20. BISHOP, W. W., and J. D. CLARK, eds. *Background to Evolution in Africa*. Univ. of Chicago Press, 1967.

21. BLANC, ALBERTO C. Some evidence for the ideologies of early man. In Washburn (302), 1961.

22. BOLWIG, NIELS. Study of the behavior of the chacma baboon. *Behavior* 14:136–63, 1959.

23. BOURLIÈRE, FRANÇOIS. *The Natural History of Mammals*. New York: Alfred A. Knopf, 1954.

24. BOYCOTT, BRIAN B. Learning in the octopus. *Scientific American*, March 1965.

25. BRADT, GLENN W. Study of beaver colonies in Michigan. *Jour. Mammalogy* 19:139–62, 1938.

26. BRAESTRUP, F. W. Study of the Arctic fox in Greenland. *Meddelelser Om Grønland* 131:1–101, 1941.

27. BRAIDWOOD, ROBERT J. The agricultural revolution. *Scientific American* Sept. 1960.

28. BRAIN, C. K. New finds at the Swartkrans Australopithecine site. *Nature* 225:1112–18, 1970.

29. BRANT, D. H., and J. L. KAVANAU. Unrewarded exploration and learning complex by wild and domestic mice. *Nature* 204:267–9, 1964.

30. BREDER, C. M., JR., and C. W. COATES. Population stability and sex ratio of *Lebistes*. *Copeia*, 1932:147–55.

31. BREY, J. R. Forest growth and glacial chronology. *Nature* 205: 440–3, 1965.

32. BROTHWELL, D. R. Evidence of early population change in Central and Southern Africa. *Man* July 1963.

33. BUCKNER, CHARLES H. Populations and ecological relationships of shrews in tamarack bogs of Manitoba. *Jour. Mammalogy* 47: 181–94, 1966.

34. BUECHNER, H. K. Territoriality as a behavioral adaptation to environment in the Uganda kob. *Proc. XVI Int. Cong. Zool.* 3:59–62, 1963.

35. BURGERS, J. M. Curiosity and play. *Science* 154:1680–1, 1966.

36. BURT, W. H. Territorial behavior and populations of some small animals in Southern Michigan. *Misc. Pub.* Museum of Zoology, Univ. of Michigan, 45, 1–58, 1940.

37. CALHOUN, J. B. Social aspects of population dynamics. *Jour. Mammalogy* 33:139–59, 1952.

38. CALHOUN, J. B. Population density and social pathology. *Scientific American* Feb. 1962.

39. CAMPBELL, BERNARD. *Human Evolution.* London: Heinemann, 1967

40. CARPENTER, C. R. *Naturalistic Behavior of Nonhuman Primates.* Univ. Park: Pennsylvania State Univ. Press, 1964. (An anthology of Carpenter's papers.)

41. CARR-SAUNDERS, SIR ALEXANDER. *The Population Problem.* Oxford: Clarendon Press, 1922.

42. CARRICK, ROBERT. Ecological significance of territory in the Australian magpie. *Proc. XIII Int. Ornithological Cong.* 740–53, 1963.

43. CARRIGHAR, SALLY. War is not in our genes. In Montagu (210).

44. CASPARI, E. W., and R. E. MARSHAK. The rise and fall of Lysenko. *Science* 149:275–8, 1965.

45. CASSIRER, ERNEST. *The Question of Jean-Jacques Rousseau,* introduction by Peter Gay. Bloomington: Indiana Univ. Press, Midland Book, 1963.

46. CHANCE, M. R. A. Attention structure as the basis of primate rank orders. *Man* Dec. 1967.

47. CHIPMAN, R. K., *et al.* Pregnancy failure in laboratory mice after multiple short-term exposure to strange male. *Nature* 210: 653, 1966.

48. CHITTY, DENNIS. Population processes in the vole and their relevance to general theory. *Canadian Jour. Zoology* 38:99–113, 1960.

49. CHOMSKY, NOAM. *Aspects of the Theory of Syntax.* Cambridge, Mass.: M.I.T. Press, 1965.

50. CHRISTIAN, J. J., and D. E. DAVIS. Endocrines, behavior and population. *Science* 146:1550–60, 1964.

51. CLULOW, F. V., and J. R. CLARKE. Pregnancy block in *Microtus agrestis* an induced ovulator. *Nature* 219:511, 1968.

52. COLEMAN, JAMES S., *et al. Equality of Educational Opportunity.* Washington: U.S. Govt. Printing Office, 1966.

53. COLLIAS, N. E. Social life and the individual among vertebrate animals. *Ann. N.Y. Acad. Sci.* 51:1074–92, 1950.

54. COSER, LEWIS A. Durkheim's conservatism. In E. Durkheim, *Essays on Sociology and Philosophy.* New York: Harper & Row, 1964.

55. COWLES, RAYMOND B. Heat induced sterility and its possible bearing on evolution. *Amer. Naturalist* 79:160–75, 1945.

56. CRAIK, K. J. W. White plumage in sea-birds. *Nature* 153:288, 1944.

57. CROOK, J. H. Two closely related weaver-bird species. *Behaviour* 21:177–232, 1963.

58. CROOK, J. H. The nature and function of territorial aggression. In Montagu (210).

59. CROOK, J. H., and J. S. GARTLAN. Evolution and primate societies. *Nature* 210:1200–3, 1966.

60. CULLEN, ESTHER. Adaptations in the kittiwake to cliff-nesting. *Ibis* 99:275–302, 1957.

61. DARLING, F. F. *A Herd of Red Deer.* London and New York: Oxford Univ. Press, 1937.

62. DARLING, F. F. Social behavior and survival. *Auk* 69:183–191, 1952.

63. DARLINGTON, C. D. *Evolution of Genetic Systems.* London: Oliver and Boyd, 1939.

64. DART, R. A. The predatory transition from ape to man. *International Anthropological and Linguistic Review*, vol. 1, no. 4, 1953.

65. DARWIN, CHARLES. *On the Origin of Species.* London: J. M. Dent, 1956. First published 1859.

66. DARWIN, CHARLES. *The Descent of Man*, 1871. Second edition, New York and London: Merrill and Baker, 1874.

67. DAVIS, D. E. Physiological analysis of aggressive behavior. In Etkin (90), 1964.

68. DAVIS, D. E. Territorial rank in starlings. *Animal Behaviour* 7:214–21, 1959.

69. DE KOCK, L. L., and A. E. ROBINSON. A lemming movement in Jämtland, Sweden, in autumn, 1963. *Jour. Mammalogy* 47:490–99, 1966.

70. DE LUMLEY, HENRY. Paleolithic camp at Nice. *Scientific American* May 1969.

71. DE VORE, IRVEN, ed. *Primate Behavior.* New York: Holt, Rinehart and Winston, 1965.

72. DE VORE, IRVEN, and K. R. L. HALL. Baboon ecology. In De Vore (71), 1965.

73. DE VORE, IRVEN, and S. L. WASHBURN. Baboon ecology and human evolution. In Howell and Bourlière (132), 1963.

74. DICKSON, D. P., *et al.* Social relationships in a herd of dairy cows. *Behaviour* 29:195–203, 1967.

75. DOBZHANSKY, TH. *Mankind Evolving.* New Haven: Yale Univ. Press, 1962.

BIBLIOGRAPHY

76. DOBZHANSKY, TH. Species after Darwin. In Barnett (14), 1958.
77. DUBOS, RENÉ. Humanistic biology. *Amer. Scientist* 53:4–19, 1965.
78. DURANT, WILL and ARIEL. *Rousseau and Revolution.* New York: Simon and Schuster, 1967.
79. EIBL-EIBESFELDT, IRENÄUS, and S. KRAMER. Ethology, the comparative study of animal behavior. *Q. Rev. Biol.* 33:181–211, 1958.
80. EIBL-EIBESFELDT, IRENÄUS. Fighting behavior of animals. *Scientific American* Dec. 1961.
81. EKLUND, CARL R. The Antarctic skua. *Scientific American* Feb. 1964.
82. ELTON, C. S. Periodic fluctuations in the numbers of animals. *Jour. Exp. Biol.* 2:119–63, 1924.
83. ERRINGTON, PAUL. Fluctuation in populations of muskrats. *Amer. Naturalist* 85:273–92, 1951.
84. ERRINGTON, PAUL. Factors limiting higher vertebrate populations. *Science* 124:304–7, 1956.
85. ESSER, ARISTIDE H., et al. Territoriality of patients on a research ward. *Recent Advances in Biological Psychiatry* 7:37–44, 1964.
86. ESSER, ARISTIDE H. Dominance hierarchy and clinical course of psychiatrically hospitalized boys. *Child Development* 39:147–57, 1968.
87. ESTES, RICHARD D. Comparative behavior of Grant's and Thomson's gazelles. *Jour. Mammalogy* 48:189–209, 1967.
88. ESTES, RICHARD D. Behavior and life history of the wildebeest. *Nature* 212:999–1000, 1966.
89. ESTES, RICHARD D. Territorial behavior of the wildebeest. Jan. 1968, in press.
90. ETKIN, WILLIAM, ed. *Social Behavior and Organization Among Vertebrates.* Chicago: Univ. of Chicago Press, 1964.
91. EWING, L. S. Fighting and death from stress in cockroach. *Science* 155:1035–6, 1967.
92. FISHER, EDNA M. Habits of the southern sea otter. *Jour. Mammalogy* 20:21–36, 1939.
93. FISHER, RONALD A. *Genetical Theory of Natural Selection.* Revised, New York: Dover, 1958. Original publication 1929.
94. FOODEN, J. Complementary specialization of male and female reproductive structure in bear macaque. *Nature* 214:939–41, 1967.
95. FRIEDAN, BETTY. *The Feminine Mystique.* New York: Norton, 1963.

96. GEORGE, WILMA. *Biologist Philosopher.* New York and London: Abelard-Schuman, 1964.

97. GLASS, B., and B. HEEZEN. Tektites and geomagnetic reversals. *Nature* 214:372, 1967.

98. GLASS, B., and B. HEEZEN. Tektites and geomagnetic reversals. *Scientific American* July 1967.

99. GOODALL, JANE VAN LAWICK. Feeding behavior of the wild chimpanzee. *Symp. Zool. Soc. London* 10:39–47, 1963.

100. GOODALL, JANE VAN LAWICK. My life among wild chimpanzees. *National Geographic* 124:272–308, Aug. 1963.

101. GOODALL, JANE VAN LAWICK. New discoveries among wild chimpanzees. *National Geographic* 128:802–31, Dec. 1965.

102. GORER, GEOFFREY. Man has no "killer" instinct. In Montagu (210), 1968.

103. GOTTESMAN, IRVING. Genetic aspects of intelligent behavior. In Norman R Ellis, ed. *Handbook of Mental Deficiency.* New York: McGraw-Hill, 1963.

104. GUARDIAN. Man an aggressive animal. London, Aug. 25, 1969.

105. GUDGER, E. W. Fishes that swim heads to tails in single file. *Copeia* 1944:152–4.

106. GUHL, A. M., and W. C. ALLEE. Measurable effects of social organization in flocks of hens. *Phys. Zool.* 17:320–47, 1944.

107. HALDANE, J. B. S. *The Causes of Evolution.* London: Longmans, 1932.

108. HALDANE, J. B. S. Quantitative measurements of rates of evolution. *Evolution* 3:51–6, 1949.

109. HALL, EDWARD T. *The Hidden Dimension.* New York: Doubleday, 1966. Also, Environmental communication, in A. H. Esser, ed., *The Use of Space by Animals and Men,* in press.

110. HALL, K. R. L. Sexual, agonistic, and derived social behavior patterns of wild chacma baboon. *Proc. Zool. Soc. London* 139:283–327, 1962.

111. HALL, K. R. L. Behavior of patas monkeys. *Folia Primatologica* 3:22–49, 1965.

112. HALL, K. R. L. Social organization of Old World monkeys and apes. *Symp. Zool. Soc. London* 14:265–89, 1965.

113. HALL, K. R. L., and IRVEN DE VORE. Baboon social behavior. In De Vore (71), 1965.

114. HALL, K. R. L., and J. S. GARTLAN. Ecology and behavior of the vervet monkey. *Proc. Zool. Soc. London* 145:37–56, 1965.

115. HARDY, ALISTER. *The Living Stream.* London: Collins, 1965.

116. HARLOW, HARRY F. Evolution of learning. In Roe and Simpson (246), 1958.

117. HARLOW, HARRY F. and MARGARET. A study of animal affection. In Southwick (275), 1963. Originally in *Natural History* 70: 48–55, 1961.

118. HARLOW, HARRY F. and MARGARET. Social deprivation in monkeys. *Scientific American* Nov. 1962.

119. HARLOW, HARRY F. and MARGARET. Affection in primates. *Discovery* Jan. 1966.

120. HARRISON, C. G. A., and F. H. CHAMALAUN. Behavior of the earth's magnetic field during a reversal. *Nature* 212:1193–5, 1966.

121. HAYS, J. D. Radiolaria in late Tertiary and Quaternary history of Antarctic seas. *Antarctic Res. Series* 5:125–184, 1965.

122. HAYS, J. D., and N. D. OPDYKE. Antarctic radiolaria, magnetic reversals, and climate change. *Science* 158:1001–11, 1967.

123. HEDIGER, H. *Psychology of Animals in Zoos and Circuses.* London: Butterworth, 1955.

124. HINDE, R. A. Biological significance of territories in birds. *Ibis* 98: 340–69, 1956.

125. HINKLE, L. E., JR., *et al.* Occupation, education, and coronary heart disease. *Science* 161:238–46, 1968.

126. HIRSCH, JERRY. Behavior genetics and individuality understood. *Science* 142:1436–42, 1963.

127. HOGAN-WARBURG, A. J. Social behavior of the ruff. *Ardea* 54:111–229, 1966.

128. HOWARD, ELIOT. *Territory in Bird Life.* London: Collins, 1948. Originally John Murray, London, 1920.

129. HOWARD, W. E., and J. T. EMLEN, JR. Intercovey social relations in the valley quail. *Wilson Bull.* 54:162–70, 1942.

130. HOWELL, F. CLARK. *Early Man.* New York: Time Inc., 1965.

131. HOWELL, F. CLARK. Remains of Hominidae . . . in the lower Omo Basin, Ethiopia. *Nature* 223:1234–9, 1969.

132. HOWELL, F. CLARK, and FRANÇOIS BOURLIÈRE, eds. *African Ecology and Human Behavior.* Chicago: Aldine, 1963.

133. HUXLEY, JULIAN. *Man in the Modern World.* New York: Harper & Row, 1939.

134. HUXLEY, JULIAN. *Evolution: A Modern Synthesis.* London: George Allen and Unwin, 1942.

135. HUXLEY, JULIAN. *Essays of a Humanist.* New York: Harper & Row, 1964.

136. IMANISHI, KINJI. Social behavior in Japanese monkeys. In Southwick (265), 1963.

137. ISAAC, GLYNN L. Traces of Pleistocene hunters. In Lee and De Vore (180), 1968.
138. ITANI, JANICHIRO. Paternal care in wild Japanese monkeys. In Southwick (265), 1963.
139. JAMES, W. The application of Theory Y at Rotterdam. *Oil and Gas Journal*, in press.
140. JAY, PHYLLIS. The Indian langur monkey. In Southwick (265), 1963.
141. JENSEN, ARTHUR H. How much can we boost IQ and scholastic achievement? *Harvard Educational Rev.* 39:1–122, 1969.
142. JORDAN, P. A., P. C. SHELTON, and D. A. ALLEN. Numbers, turnover and social structure of the Isle Royale wolf population. *Amer. Zoology* 7:233–52, 1967.
143. JUNG, C. G. *Psychology of the Unconscious*. London: Kegan Paul, 1919.
144. KAUFMAN, JOHN H. Social relations of males in a free ranging band of rhesus monkeys. In Altmann (8), 1967.
145. KAVANAU, J. LEE. Behavior of captive white-footed mouse. *Science* 155:1623–39, 1967.
146. KAWAMURA, SYUNZA. Process of sub-culture propagation among Japanese monkeys. In Southwick (265), 1963.
147. KEITH, ARTHUR. *A New Theory of Human Evolution*. London: Watts, 1948.
148. KENYATTA, JOMO. *Facing Mount Kenya*. London: Martin Secker & Warburg, 1938. Mercury Books, 1961.
149. KERN, JAMES A. Observations on the habits of the proboscis monkey. *Zoologica* 49:183–92, 1964.
150. KING, JOHN A. Social behavior . . . and population dynamics in a black-tailed prairie-dog town. *Contributions from the Laboratory of Vert. Biol.*, Univ. of Michigan, 67:1–123, 1955.
151. KINZEL, AUGUSTUS. Body-buffer zones in violent prisoners. *Amer. Jour. Psychiatry*, 1970, in press.
152. KOESTLER, ARTHUR. The predicament of modern man. In Ng (226), 1968.
153. KOFORD, CARL B. Rank of mothers and sons in bands of rhesus monkeys. *Science* 141:356–7, 1963.
154. KORN, N., and F. THOMPSON, eds. *Human Evolution*. New York: Holt, Rinehart & Winston, 1967.
155. KORTLANDT, A. *Aspects and Prospects of the Concept of Instinct*. Leyden: Brill, 1955.
156. KORTLANDT, A. Chimpanzees in the wild. *Scientific American* May 1962.

157. KORTLANDT, A., and M. KOOIJ. Protohominid behavior in primates. *Symp. Zool. Soc. London* 10:61–88, 1963.

158. KREBS, C. J., and K. T. DELONG. A *microtus* population with supplemental food. *Jour. Mammalogy* 46:566–73, 1965.

159. KRUUK, HANS. Clan system and feeding habits of spotted hyenas. *Nature* 209:1257–8, 1966.

160. KRUUK, HANS. Hyenas, the hunters nobody knows. *National Geographic* 134:44–57, July 1968.

161. KUHME, WOLFDIETRICH. Communal food distribution and division of labor in African hunting dogs. *Nature* 205:443–4, 1965.

162. KUMMER, H., and F. KURT. Social units of a free-living population of hamadryas baboons. *Folia Primatologica* 1:4–19, 1963.

163. KURILOFF, A. H. Experiment in management. *Personnel* Nov.–Dec. 1963.

164. KURILOFF, A. H. *Reality in Management.* New York: McGraw-Hill, 1966.

165. LACK, DAVID. *The Life of the Robin.* London: Pelican, 1953. Originally published 1943.

166. LACK, DAVID. *Natural Regulation of Animal Numbers.* Oxford: Clarendon Press, 1954.

167. LACK, DAVID. Significance of clutch-size in swift and grouse. (With Wynne-Edwards' reply.) *Nature* 203:98–9, 1964.

168. LACK, DAVID, *et al.* Survival in relation to brood size in tits. *Proc. Zool. Soc. London* 128:313–26, 1957.

169. LANCASTER, JANE B., and R. B. LEE. Annual reproductive cycles in monkeys and apes. In De Vore (71), 1965.

170. LAUGHLIN, WILLIAM S. Hunting: an integrating biobehavior system and its evolutionary significance. In Lee and De Vore (180), 1968.

171. LAWS, R.M., and I. S. C. PARKER. Recent studies of elephant populations in East Africa. *Symp. Zool. Soc. London* 21:319–59, 1968.

172. LEAKEY, L. S. B. A new Lower Pliocene fossil primate from Kenya. *Ann. & Magazine Nat. Hist.,* Ser XIII, 4:689–96, 1961.

173. LEAKEY, L. S. B. *Olduvai Gorge 1951–1961.* Cambridge Univ. Press, 1965.

174. LEAKEY, L. S. B. Notes on the mammalian faunas of East Africa. In Bishop and Clark (20), 1967.

175. LEAKEY, L. S. B. Bone-smashing by Late Miocene hominid. *Nature* 218:528–30, 1968.

176. LEAKEY, L. S. B., P. V. TOBIAS, and J. R. NAPIER. A new species of genus Homo from Olduvai Gorge. *Nature* 202:7–9, 1964.

427

177. LEAKEY, L. S. B. and MARY. Recent discoveries in Tanganyika. *Nature* 202:5–7, 1964.
178. LEAKEY, MARY. Review of the Olduwan culture from Olduvai Gorge. *Nature* 210:462–6, 1966.
179. LEE, RICHARD B. What hunters do for a living. In Lee and De Vore (180), 1968.
180. LEE, RICHARD B., and IRVEN DE VORE, eds. *Man the Hunter*. Chicago: Aldine, 1968.
181. LE GROS CLARK, WILFRED. *Man-Apes or Ape-Men?* New York: Holt, Rinehart & Winston, 1967.
182. LENNEBERG, ERIC H. *Biological Foundations of Language*. New York: Wiley, 1967.
183. LÉVI-STRAUSS, CLAUDE. Social and psychological aspects of chieftainship in a primitive tribe. In *Comparative Political Systems*, R. Cohen and J. Middleton, eds., Garden City, N.Y.: Natural History Press, 1967.
184. LEYHAUSEN, PAUL. The sane community and density problem. *Discovery* Sept. 1965.
185. LEYHAUSEN, PAUL. Communal organization of solitary animals. *Symp. Zool. Soc. London* 14:249–63, 1965.
186. LINDLEY, MARY. Keep off the grass. *Nature* 216:1166–7, 1967.
187. LIVINGSTONE, FRANK B. On the non-existence of human races. In Korn and Thompson (154), 1967.
188. LORENZ, KONRAD Z. The companion in the bird's world. *Auk* 54: 245–73, 1937.
189. LORENZ, KONRAD Z. *King Solomon's Ring*. New York: Crowell, 1952.
190. LORENZ, KONRAD Z. *On Aggression*. London: Methuen, 1966.
191. LOVERIDGE, ARTHUR. Cannibalism in the common coral snake. *Copeia* 1944:254.
192. MALTHUS, THOMAS. *Population*. New York: Modern Library.
193. MARAIS, EUGÈNE. Baboons, hypnosis and insanity. *Psyche* 7:104–10, 1926.
194. MARAIS, EUGÈNE. *The Soul of the Ape*. New York: Atheneum, 1969.
195. MARTIN, MALACHI. *The Encounter*. New York: Farrar, Straus and Giroux, 1969.
196. MASLOW, A. H. *Motivation and Personality*. New York: Harper and Row, 1954.
197. MASLOW, A. H. Synergy in the society and in the individual. *Jour. Individ. Psych.* 20:153–64, 1964.
198. MASON, WILLIAM A. Effects of environmental restriction on the so-

cial development of rhesus monkeys. In Southwick (275), 1963. Originally published 1958.

199. MAST, S. O., and L. C. PUSCH. Modification of response in the amoeba. *Biol. Bull.* 46:55–9, 1924.

200. MASTERS, ROGER D. *The Political Philosophy of Rousseau.* Princeton Univ. Press, 1968.

201. MASURE, R. H., and W. C. ALLEE. Social order in the . . . chicken and pigeon. *Auk* 51:306–27, 1934.

202. MATHER, KENNETH. *Human Diversity.* London: Oliver & Boyd, 1964.

203. MATHER, KENNETH. Variation and selection of polygenic characters. *Jour. Genetics* 41:159–93, 1941.

204. MATTHEWS, G. V. T. Orientation of untrained pigeons. *Jour. Exp. Biol.* 30:268–76, 1953.

205. MAYR, ERNST. *Animal Species and Evolution.* Cambridge, Mass.: Harvard Univ. Press (Belknap), 1963.

206. MAYR, ERNST. Behavior and systematics. In Roe and Simpson (246), 1958.

207. MAYR, ERNST. Speciation phenomenon in birds. *Amer. Naturalist* 74:249–78, 1940.

208. MOE, HENRY ALLEN. On the need for an aristocracy. Washington, D.C.: Cosmos Club, 1965.

209. MOFFAT, C. B. Spring rivalry of birds. *Irish Naturalist* 12:152–66, 1903.

210. MONTAGU, M. F. ASHLEY. *Man and Aggression.* New York: Oxford Univ. Press, 1968.

211. MORRIS, DESMOND. *The Biology of Art.* London: Methuen, 1962.

212. MORRIS, DESMOND. *The Human Zoo.* New York: McGraw-Hill, 1969.

213. MORRIS, DESMOND. Homosexuality in the ten-spined stickleback. *Behaviour* 4:233–61, 1952.

214. MURPHY, ROBERT F. Social distance and the veil. *Amer. Anthropologist* 66:1257–74, 1964.

215. MYERS, K. Morphological changes in the adrenal glands of wild rabbits. *Nature* 213:147–50, 1967.

216. MYKYTOWYCZ, R. Territoriality in rabbit populations. *Australian Nat. Hist.* 14:326–9, 1964.

217. MC BRIDE, GLEN. The conflict of overcrowding. *Discovery* Apr. 1966.

218. MC GREGOR, DOUGLAS. *The Human Side of Enterprise.* New York: McGraw-Hill, 1960.

219. MAC LEAN, PAUL D. Psychosomatic disease and the "Visceral Brain." *Psychosomatic Medicine* 11:338–53, 1949.
220. MAC LEAN, PAUL D. Man and his animal brains. *Modern Medicine* Feb. 3, 1964.
221. MAC LEAN, PAUL D. Alternative neural pathways to violence. In Ng (226), 1968.
222. MAC LULICH, D. A. The varying hare. *Univ. Toronto Biol. Series* 43:1–136, 1937.
223. MC MAHON, B. E., *et al.* Kiaman magnetic interval in Western U.S. *Science* 155:1012–3, 1967.
224. NAPIER, JOHN. The antiquity of human walking. *Scientific American* Apr. 1967.
225. NAPIER, JOHN. Five steps to man. *Discovery* June 1964.
226. NG, LARRY, ed. *Alternatives to Violence.* New York: Time-Life Books, 1968.
227. NICE, M. M., and J. TER POLKWYK. Enemy recognition by the song sparrow. *Auk* 58:195–214, 1941.
228. NOBLE, G. K. Sexual selection among fishes. *Biol. Reviews* 13: 133–58, 1938.
229. OAKLEY, K. P. *Man the Tool-maker.* London: British Museum, 1949.
230. OAKLEY, K. P. *Frameworks for Dating Fossil Man.* Chicago: Aldine, 3rd edition, 1968.
231. OAKLEY, K. P. Of man's use of fire . . . tool-making, hunting. In Washburn (302), 1962.
232. OLIVER, ROLAND. *History of East Africa.* Oxford Univ. Press, 1963.
233. PALMGREN, PONTUS. Short-term fluctuations of numbers in northern birds and mammals. *Oikos* 1:114–21, 1949.
234. PERRINS, C. M. Population fluctuations and clutch-size in great tits. *Jour. Anim. Ecol.* 34:601, 1965.
235. PETTER, J.-J. *L'Ecologie et l'Ethologie des Lémuriens Malgaches.* Mémoires du Muséum National d'Histoire Naturelle, Tome XXVII, Fascicule 1, 1962.
236. PHILLIPS, W. W. A. Observations of nesting of the Ceylon black-backed pied shrike. *Ibis* 4 (ser. 14): 450–4, 1940.
237. PILBEAM, DAVID. Gigantopithecus and the Origins of Hominidae. *Nature* 225:516–18, 1970.
238. PITTENDRIGH, COLIN S. Adaptation, natural selection, and behavior. In Roe and Simpson (246), 1958.
239. POSNANSKY, M. Bantu genesis. *Uganda Jour.* 25:86–93, 1961.
240. READ, CARVETH. *The Origin of Man.* Cambridge Univ. Press, 1925.

241. RENSCH, BERNARD. Increase of learning capability with increase of brain size. *Amer. Naturalist* 90:81–95, 1956.

242. REYNOLDS, V. Open groups in hominid evolution. *Man* 1:441–52, 1966.

243. REYNOLDS, V. Behavior and social organization of forest chimpan ███. *Folia Primatologica* 1:95–102, 1963.

244. RIPLEY, SUZANNE. Pattern of socialization in Ceylon langurs. *Amer. Anthropological Soc. Symp.* Nov. 23, 1963.

245. RIPLEY, SUZANNE. Intergroup encounters among Ceylon gray lan gurs. In Altmann (8), 1967.

246. ROE, ANNE, and G. G. SIMPSON, eds. *Behavior and Evolution.* New Haven, Conn.: Yale Univ. Press, 1958.

247. ROEDER, K. D. Sexual behavior of the praying mantis. *Biol. Bull.* 69:203–20, 1935.

248. ROEDER, K. D. Control of tonus and locomotor activity in the pray ing mantis. *Jour. Exp. Zool.* 76:353–74, 1937.

249. ROMER, ALFRED S. Major steps in vertebrate evolution. *Science* 158:1629–37, 1967.

250. ROUSSEAU, JEAN-JACQUES. *The Social Contract.* New York: Hafner, 1962. Originally published 1762.

251. ROUSSEAU, JEAN-JACQUES. *First and Second Discourses,* ed. Roger D. Masters. New York: St. Martin's Press, 1964. Originally published 1750, 1755.

252. ROUSSEAU, JEAN-JACQUES. *Emile.* Paris: Garnier-Flammarion, 1966. Originally published 1762.

253. ROUSSEAU, JEAN-JACQUES. *The Confessions.* London: Penguin, 1953. Completed 1765, originally published 1781.

254. RUSSELL, CLAIRE and W. M. S. *Violence, Monkeys and Men.* Lon don: Macmillan, 1968.

255. SACKETT, GENE P. Monkeys reared in isolation with pictures as vis ual input: evidence for IRM. *Science* 154:1468–73, 1966.

256. SAHLINS, MARSHALL D. The origins of society. *Scientific American* Sept. 1960.

257. SCHALLER, GEORGE B. *The Mountain Gorilla.* Univ. of Chicago Press, 1963.

258. SCHALLER, GEORGE B. *The Deer and the Tiger.* Univ. of Chicago Press, 1967.

259. SCHALLER, GEORGE B. Life with the king of the beasts. *National Geographic* 135:494–519, 1969.

260. SCHALLER, GEORGE B., and GORDON R. LOWTHER. The relevance of carnivore behavior to the study of early hominids. *Southwest ern Jour. Anthropology* 25:307–41, 1969.

261. SCHENKEL, R. Ausdrucksstudien an Wölfen. *Behaviour* 1:81–129, 1947.

262. SCOTT, JOHN W. Mating behavior of the sage grouse. *Auk* 59: 477–98, 1942.

263. SEBEOK, THOMAS A. Animal communication. *Science* 147:1006–14, 1965.

264. SHAW, BERNARD. *Androcles and the Lion.* New York: Dodd, Mead.

265. SILBERBAUER, G. *Bushman Survey, 1st and 2nd reports.* Mafeking: Bechuanaland Secretariat, 1960–1961.

266. SIMONS, ELWYN L. The early relatives of man. *Scientific American* July 1964.

267. SIMONS, ELWYN L. Late Miocene Hominid from Fort Ternan, Kenya. *Nature* 221:448–51, 1969.

268. SIMONS, ELWYN L., and P. C. ETTEL. Gigantopithecus. *Scientific American* Jan. 1970.

269. SIMPSON, G. G. Behavior and evolution. In Roe and Simpson (246), 1958.

270. SIMPSON, G. G. *The Meaning of Evolution.* New Haven, Conn. Yale Univ. Press, 1951.

271. SKINNER, B. F. The phylogeny and ontogeny of behavior. *Science* 153:1205–13, 1966.

272. SKINNER, B. F. *Science and Human Behavior.* New York: Macmillan, 1953.

273. SOLOMON, M. E. Natural control of animal numbers. *Jour. Anim. Ecol.* 18:1–35, 1949.

274. SOMMER, ROBERT. Studies in personal space. *Sociometry* 22: 247–60, 1959.

275. SOUTHWICK, C. H., ed. *Primate Social Behavior.* New York: Van Nostrand, 1963.

276. SOUTHWICK, C. H. Patterns of intergroup social behavior in primates. *Ann. N.Y. Acad. Sci.* 102:436–54, 1962.

277. SOUTHWICK, C. H. Experimental study of intragroup agonistic behavior in rhesus monkey. *Behaviour* 28:182–209, 1967.

278. STEWARD, JULIAN H. Scientific responsibility in modern life. *Science* 159:147–8, 1968.

279. STORR, ANTHONY. *Human Aggression.* New York: Atheneum, 1968.

280. STRUHSAKER, THOMAS T. Social structure among vervet monkeys. *Behaviour* 29:83–120, 1967.

281. STRUHSAKER, THOMAS T. Behavior of vervet monkeys and other Cercopithecines. *Science* 156:1197–1203, 1967.

282. SUGIYAMA, YUKIMARU. Social organization of Hanuman langurs. In Altmann (8), 1967.

BIBLIOGRAPHY

283. SUTTLES, GERALD B. *The Social Order of the Slum.* Univ. of Chicago Press, 1968.

284. TATTERSALL, IAN. Ecology of north Indian Ramapithecus. *Nature* 221:451–2, 1969.

285. TAX, SOL, ed. *Horizons of Anthropology.* London: Allen & Unwin, 1964

286. THODAY, J. M. Components of fitness. *Symp. Soc. Exp. Biology* 7:96–113, 1953.

287. THOMA, A. L'Occipital de l'Homme Mindélien de Vértesszöllös. *L'Anthropologies* (Paris) 70:495–534, 1966.

288. THOMAS, J. W., *et al.* Social behavior in a white-tailed deer herd containing hypogonadal males. *Jour. Mammalogy* 46:314–27, 1965.

289. THORPE, W. H. *Learning and Instinct in Animals.* London: Methuen, 1956. Revised 1962.

290. TIGER, LIONEL. *Men in Groups.* New York: Random House, 1969.

291. TINBERGEN, N. *The Herring Gull's World.* London: Collins, 1953.

292. TINBERGEN, N. *The Study of Instinct.* Oxford: Clarendon Press, 1951.

293. TINBERGEN, N. *Social Behaviour in Animals.* London: Methuen, 1953.

294. TIXIER, J. Ensembles industriels . . . dans l'Afrique du Nord-Ouest. In Bishop and Clark (20), 1967.

295. TOBIAS, P. V. Cranial capacity in anthropoid apes, australopithecus, and Homo habilis. *South African Jour. Sci.* 64:81–91, 1968.

296. TURNBULL, COLIN M. The importance of flux in two hunting societies. In Lee and De Vore (180), 1968.

297. UFFEN, ROBERT J. Influence of earth's core on origin and evolution of life. *Nature* 198:143–4, 1963.

298. VALLOIS, HENRI. Evidence of skeletons. In Washburn (302), 1962.

299. WADDINGTON, C. H. Theories of evolution. In Barnett (14), 1958.

300. WADDINGTON, C. J. Paleomagnetic field reversals and cosmic radiation. *Science* 158:913–15, 1967.

301. WALLACE, ALFRED R. Origin of human races and the antiquity of man . . . *Jour. Anthropological Soc. London,* 1864: clviii.

302. WASHBURN, S. L., ed. *Social Life of Early Man.* London: Methuen, 1962.

303. WASHBURN, S. L., and VIRGINIA AVIS. Evolution of Human Behavior. In Roe and Simpson (246), 1958.

304. WASHBURN, S. L., and IRVEN DE VORE. Social life of baboons, *Scientific American* June 1961.

433

305. WASHBURN, S. L., and D. HAMBURG. Implications of primate research. In De Vore (71), 1965.
306. WASHBURN, S. L., and C. S. LANCASTER. The evolution of hunting. In Korn and Thompson (154), 1967.
307. WATSON, ADAM. Population control by territorial behavior in red grouse. *Nature* 215:1274–5, 1967.
308. WATSON, JAMES D. *The Double Helix.* New York: Atheneum, 1968.
309. WECKER, STANLEY C. Habitat selection. *Scientific American* Oct. 1964.
310. WHITE, THEODORE H. *Caesar at the Rubicon.* New York: Atheneum, 1968.
311. WHYTE, L. L. *Internal Factors in Evolution.* London: Tavistock, 1965.
312. WHYTE, WILLIAM H. *The Organization Man.* New York: Simon and Schuster, 1956.
313. WICKLER, WOLFGANG. *Mimicry in Plants and Animals.* London: Weidenfeld and Nicolson, 1968.
314. WIESENFELD, S. L. Sickle cell trait in human biological and cultural evolution. *Science* 157:1134–40, 1967.
315. WILSON, E. O. Competitive and aggressive behavior. Smithsonian symposium, May, 1969.
316. WILSON, E. O., et al. The first mesozoic ant. *Science* 157:1038–40, 1967.
317. WOODBURN, JAMES. An introduction to Hazda ecology. In Lee and De Vore (180), 1968.
318. WRIGHT, SEWALL. Evolution in Mendelian populations. *Genetics* 16:97–159, 1931.
319. WRIGHT, SEWALL. Modes of selection. *Amer. Naturalist* 90:5–24, 1956.
320. WYNNE-EDWARDS, V. C. *Animal Dispersion in Relation to Social Behavior.* London: Oliver & Boyd, 1962.
321. WYNNE-EDWARDS, V. C. Population control in animals. *Scientific American* Aug. 1964.
322. ZIGLER, EDWARD. Familial mental retardation: a continuing dilemma. *Science* 155:292–8, 1967.

Index

L